ACCIDENTAL
EMPIRES

ACCIDENTAL EMPIRES

HOW THE BOYS OF SILICON VALLEY
MAKE THEIR MILLIONS,
BATTLE FOREIGN COMPETITION,
AND STILL CAN'T GET A DATE

ROBERT X. CRINGELY

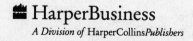

HarperBusiness
A Division of HarperCollinsPublishers

A hardcover edition of this book was published in 1992 by Addison-Wesley Publishing Company, Inc. It is here reprinted by arrangement with Addison-Wesley Publishing Company, Inc.

HarperCollins books may be purchased for educational, business, or sales promotional use. For information please write: Special Markets Department, HarperCollins Publishers, Inc., 10 East 53rd Street, New York, NY 10022.

First HarperBusiness edition published 1993. Reissued 1996.

Designed by Janis Owens

The Library of Congress has catalogued the previous edition as follows:
Cringely, Robert X.
    Accidental empires: how the boys of Silicon Valley make their millions, battle foreign competition, and still can't get a date / Robert X. Cringely.
        p. cm.
    Includes bibliographical references.
    ISBN 0-88730-621-7
    1. Computer industry—California—Santa Clara County. I. Title.
HD9696.C63U51586 1993
338.4'7004'0979473—dc20   92-53427

ISBN 0-88730-855-4
        98 99 00 RRD 10 9 8

For Pammy, who knows we need the money

CONTENTS

PREFACE TO THE
1996 EDITION

In his novel *Brighton Rock*, Graham Greene's protagonist, a cocky 14-year-old gang leader named Pinky, has his first sexual experience. Nervously undressing, Pinky is relieved when the girl doesn't laugh at the sight of his adolescent body. I know exactly how Pinky felt.

When I finished writing this book five years ago, I had no idea how it would be received. Nothing quite like it had been written before. Books about the personal computer industry at that time either were mired in technobabble or described a gee-whiz culture in which there were no bad guys. In this book, there are bad guys. The book contained the total wisdom of my fifteen-plus years in the personal computer business. But what if I had no wisdom? What if I was wrong?

With this new edition, I can happily report that the verdict is in: for the most part, I was right. Hundreds of thousands of readers, many of whom work in the personal computer industry, have generally validated the material presented here. With the exception of an occasional typographical error and my stupid prediction that Bill Gates would not marry, what you are about to read is generally accepted as right on the money.

Not that everyone is happy with me. Certainly Bill Gates doesn't like to be characterized as a megalomaniac, and Steve Jobs doesn't like to be described as a sociopath, but that's what they are. Trust me.

This new edition is prompted by a three-hour television miniseries based on the book and scheduled to play during 1996 in most of the English-speaking world. The production, which took a year to make, includes more than 120 hours of interviews with the really important people in this story—even the megalomaniacs and sociopaths. These interviews, too, confirmed many of the ideas I originally presented in the book, as well as providing material for the new chapters at the end.

What follows are the fifteen original chapters from the 1992 edition and a pair of new ones updating the story through early 1996.

So let the computer chips fall where they may.

PREFACE

The woman of my dreams once landed a job as the girls' English teacher at the Hebrew Institute of Santa Clara. Despite the fact that it was a very small operation, her students (about eight of them) decided to produce a school newspaper, which they generally filled with gossipy stories about each other. The premiere issue was printed on good stock with lots of extra copies for grandparents and for interested bystanders like me. The girls read the stories about each other, then read the stories about each other *to* each other, pretending that they'd never heard the stories before, much less written them. My cats do something like that, too, I've noticed, when they hide a rubber band under the edge of the rug and then allow themselves to discover it a moment later.

The newspaper was a tremendous success until mid-morning, when the principal, Rabbi Porter, finally got around to reading his copy. "Where," he asked, "are the morals? None of these stories have morals!"

I've just gone through this book you are about to read, and danged if I can't find a moral in there either. Just more proof, I guess, of my own lack of morality.

There are lots of people who aren't going to like this book, whether they are into morals or not. I figure there are three distinct groups of people who'll hate this thing.

Hate group number one consists of most of the people who are mentioned in the book.

Hate group number two consists of all the people who *aren't* mentioned in the book and are pissed at not being able to join hate group number one.

Hate group number three doesn't give a damn about the other two hate groups and will just hate the book because somewhere I write that object-oriented programming was invented in Norway in 1967, when they know it was invented in *Bergen*, Norway, on a rainy afternoon in late 1966. I never have been able to please these folks, who are mainly programmers and engineers, but I take some consolation in knowing that there are only a couple hundred thousand of them.

My guess is that most people won't hate this book, but if they do, I can take it. That's my job.

Even a flawed book like this one takes the cooperation of a lot of really flawed people. More than 200 of these people are personal computer industry veterans who talked to me on, off, or near the record, sometimes risking their jobs to do so. I am especially grateful to the brave souls who allowed me to use their names.

The delightfully flawed reporters of *InfoWorld*, who do most of my work for me, continued to pull that duty for this book, too, especially Laurie Flynn, Ed Foster, Stuart Johnston, Alice LaPlante, and Ed Scannell.

A stream of *InfoWorld* editors and publishers came and went during the time it took me to research and write the book. That they allowed me to do it in the first place is a miracle I attribute to Jonathan Sacks.

Ella Wolfe, who used to work for Stalin and knows a lost

cause when she sees one, faithfully kept my mailbox overflowing with helpful clippings from the *New York Times*.

Paulina Borsook read the early drafts, offering constructive criticism and even more constructive assurance that, yes, there was a book in there someplace. Maybe.

William Patrick of Addison-Wesley believed in the book even when he didn't believe in the words I happened to be writing. If the book has value, it is probably due to his patience and guidance.

For inspiration and understanding, I was never let down by Pammy, the woman of my dreams.

Finally, any errors in the text are mine. I'm sure you'll find them.

ACCIDENTAL

EMPIRES

▸ ▸ ▸ ▸ ▸ ▸ ▸ ▸ ▸ ▸ ▸ ▸

THE DEMO-GOD

Years ago, when you were a kid and I was a kid, something changed in America. One moment we were players of baseball, voters, readers of books, makers of dinner, arguers. And a second later, and for every other second since then, we were all just shoppers.

Shopping is what we do; it's entertainment. Consumers are what we are; we go shopping for fun. Nearly all of our energy goes into buying—thinking about what we would like to buy or earning money to pay for what we have already bought.

We invented credit cards, suburban shopping malls, and day care just to make our consumerism more efficient. We sent our wives, husbands, children, and grandparents out to work, just to pay for all the stuff we wanted—needed—to buy. We invented a thousand colors of eye shadow and more than 400 different models of automobiles, and forced every garage band in America to make a recording of "Louie Louie," just so we'd have enough goods to choose between to fill what free time remained. And when, as Americans are wont to do, we surprised ourselves by coming up with a few extra dollars and a few extra hours to spare, we invented entirely new classes of consumer products to satisfy our addiction. Why else would anyone spend $19.95 to buy an Abdomenizer exercise machine?

I blame it all on the personal computer.

Think about it for a moment. Personal computers came along in the late 1970s and by the mid-1980s had invaded every office and infected many homes. In addition to being the ultimate item of conspicuous consumption for those of us who don't collect fine art, PCs killed the office typewriter, made most secretaries obsolete, and made it possible for a 27-year-old M.B.A. with a PC, a spreadsheet program, and three pieces of questionable data to talk his bosses into looting the company pension plan and doing a leveraged buy-out.

Without personal computers, there would have been— could have been—no Michael Milkens or Ivan Boeskys. Without personal computers, there would have been no supply-side economics. But, with the development of personal computers, for the first time in history, a single person could gather together and get a shaky handle on enough data to cure a disease or destroy a career. Personal computers made it possible for businesses to move further and faster than they ever had before, creating untold wealth that we had to spend on *something*, so we all became shoppers.

Personal computers both created the longest continuous peacetime economic expansion in U.S. history and ended it.

Along the way, personal computers themselves turned into a very big business. In 1990, $70 billion worth of personal computer hardware and software were sold worldwide. After automobiles, energy production, and illegal drugs, personal computers are the largest manufacturing industry in the world and one of the great success stories for American business.

And I'm here to tell you three things:

1. It all happened more or less by accident.

4

2. The people who made it happen were amateurs.
3. And for the most part they still are.

Several hundred users of Apple Macintosh computers gathered one night in 1988 in an auditorium in Ann Arbor, Michigan, to watch a sneak preview demonstration of a new word processing application. This was consumerism in its most pure form: it drew potential buyers together to see a demonstration of a product they could all use but *wouldn't be allowed to buy*. There were no boxes for sale in the back of the room, no "send no money, we'll bill you later." This product flat wasn't for sale and wouldn't be for another five months.

Why demonstrate it at all? The idea was to keep all these folks, and the thousands of people they would talk to in the coming weeks, from buying some competitor's program before this product—this Microsoft Word 3.0—was ready for the market. Macintosh users are the snobs of the personal computer business. "Don't buy MacWrite II, WordPerfect for Macintosh, or Write-Now," they'd urge their friends and co-workers. "You've got to wait for Microsoft Word 3.0. It's radical!"

But it also didn't work.

To make the demonstration even more compelling, it was to be given by Bill Gates, Microsoft's billionaire boy chairman of the board who had flown in from Seattle for that night only. (This follows the theory that if Chrysler issued invitations to look through a telescope at one of its new minivans circling a test track, more people would be willing to look if Lee Iacocca was the driver.)

There is an art to demonstrating a computer program like this—a program that isn't really finished being written. The major parts of the program were there, but if the software had been complete, Microsoft would have been taking money for it. It *would* have been for sale in the back of the room. The fact that this was only a demonstration and that the only fingers touching the keyboard that night would be those of the highly talented Bill Gates proved that the program was in no way ready to be let loose among paying customers.

What the computer users would be seeing was not really a demonstration of software but a virtuoso performance of man and machine. Think of Microsoft Word 3.0 as a minefield in Kuwait and Bill Gates as a realtor trying to sell a few lots there before all of the land mines have been cleared. To show how safe the property is, he'd give a tour, steering prospects gently away from the remaining mines without telling them they were even in danger.

"Looks safe to me, honey," the prospective buyer would say. "Let's talk business while the kids play in the yard."

"NO!!!"

That night in Ann Arbor, according to testers back in the Microsoft quality assurance department, the version of Microsoft Word that Gates was demonstrating contained six land mines. There were known to be six Type-A bugs in the software, any one of which could lock up the Macintosh computer in an instant, sending Aunt Helen's gothic romance into the ether at the same time. All Gates had to do was guide his demo past these six danger areas to make Ann Arbor and the rest of the Macintosh world think that all was well with Microsoft Word 3.0.

Gates made it through the demonstration with only one mistake that completely locked up—crashed—the computer. Not good enough for the automotive world, of course, where having to push the car back from the test drive would usually kill a sale, but computer users are forgiving souls; they don't seem to mind much if the gas tank of their digital Pinto occasionally explodes. Heck, what's one crash among friends?

In fact, the demo was brilliant, given that the Microsoft QA department had no idea how bad the program really was. Word 3.0 turned out to have not six but more than 600 major bugs when it finally shipped five months later, proving once again that Bill Gates is a demo-god.

Late night in Ann Arbor brings with it the limited pleasures of any college town—movie houses, pizzerias, and bars, each filled with a mix of students and townies that varies in direct relation to its distance from the University of Michigan campus. Bored with the Lysol ambiance of the Holiday Inn, the pair aimed their rental car into the heart of town, looking for something, well, different. Bill Gates sat on the passenger side, sniffing like a setter the evening air through his open window, a 33-year-old billionaire on the prowl.

The Word 3.0 demo was over, but Gates, now a little drunk, apparently had a few things left to prove.

"Here, stop here!" Gates commanded, jumping unsteadily from the car as it settled next to the curb near a group of young blacks.

"What's happening!" the pencil-necked billionaire cheerfully greeted the assembled boom boxers, who clearly had no idea who or what he was—this bespectacled white boy with greasy blond hair and bratwurst skin, wearing a blue and white plaid polyester shirt and green pullover sweater.

"Bill, let's go someplace else," called Gates's companion from the driver's seat.

"Yeah, Bill, go someplace else," said one of the young blacks.

"Nah, I want to rap. I can talk to these guys, you'll see!"

This is not just a gratuitous "Bill Gates gets drunk" story. "I can [fill in the blank], you'll see!" is the battle cry of the personal computing revolution and the entire philosophical basis of Microsoft's success and Gates's $4 billion fortune.

This guy thinks he has something to prove. A zillion dollars isn't enough, 7,000 employees who idolize him aren't enough—in fact, *nothing* is enough to prove to Bill Gates and to all the folks like him in the personal computer business that they are finally safe from the bigger, stronger, stupider kids who used to push them around on the playground.

"I can [fill in the blank], you'll see!" is a cry of adolescent defiance and enthusiasm, a cry as much against the status quo as it is in favor of something new. It's a cry at once of confidence and of the uncertainty that lies behind any overt need to prove one's manhood. And it's the cry that rings, at least metaphorically, across the desks of 45 million Americans as they power up their personal computers at the start of each working day.

There was no urge to fly, to see the world, to win a war, to cure disease, or even to get rich that explains how the personal computer business came to be or even how it runs today. Instead, the game was started to satisfy the needs of disenfranchised nerds like Bill Gates who didn't meet the macho standards of American maleness and so looked for a way to create their own adolescent alternative to the adult world and, through that creation, gain the admiration of their peers.

This is key: they did it (and do it) to impress each other.

In the mid-1970s, when it was hard to argue that there even was a PC industry, 19-year-old Bill Gates thought that he could write a high-level programming language—a version of the BASIC language—to run on the then-unique Altair hobbyist computer. Even the Altair's designers thought that their machine was too primitive to support such a language, but Gates, with his friend Paul Allen, thought otherwise. "We can write that BASIC interpreter, you'll see!" they said. And they were right: Microsoft was born.

When Steve Wozniak built the first Apple computer, his goal was not to create an industry, to get rich, or even to produce more than one of the machines; he just wanted to impress his friends in Silicon Valley's Homebrew Computer Club. The idea to manufacture the Apple I for sale came from Wozniak's friend, Steve Jobs, who wanted to make his mark too, but lacked Woz's technical ability. Offering a VW Microbus and use of his parents' garage in payment for a share of his friend's glory, Jobs literally created the PC industry we know today.

These pioneers of personal computing were people who had little previous work experience and no previous success. Wozniak was an undistinguished engineer at Hewlett-Packard. Jobs worked part time at a video game company. Neither had graduated from college. Bill Gates started Microsoft after dropping out of Harvard during his sophomore year. They were just smart kids who came up with an angle that they have exploited to the max.

The Airport Kid was what they called a boy who ran errands and did odd jobs around a landing field in exchange for airplane rides and the distant prospect of learning to fly. From Lindbergh's day on, every landing strip anywhere in America had such a kid— sometimes several—who'd caught on to the wonder of flight and wasn't about to let go.

Technologies usually fade in popularity as they are replaced by new ways of doing things, so the lure of flight must have been awesome, because the airport kids stuck around America for generations. They finally disappeared in the 1970s, killed not by a transcendant technology but by the dismal economics of flight.

The numbers said that unless all of us were airport kids, there would not be economies of scale to make flying cheap enough for any of us. The kids would never own their means of flight. Rather than live and work in the sky, they could only hope for an occasional visit. It was the final understanding of this truth that killed their dream.

When I came to California in 1977, I literally bumped into the Silicon Valley equivalent of the airport kids. They were teenagers, mad for digital electronics and the idea of building their own computers. We met diving through dumpsters behind electronics factories in Palo Alto and Mountain View, looking for usable components in the trash.

But where the airport kids had drawn pictures of airplanes in their school notebooks and dreamed of learning to fly, these new kids in California actually built their simple computers and taught themselves to program. In many ways, their task was easier, since they lived in the shadow of Hewlett-Packard and the semiconductor companies that were rapidly filling what had come to be called Silicon Valley. Their parents often worked in the electronics industry and recognized its value. And unlike flying, the world of microcomputing did not require a license.

Today there are 45 million personal computers in America. Those dumpster kids are grown and occupy important positions in computer and software companies worth billions of dollars. Unlike the long-gone airport kids, these computer kids came to control the means of producing their dreams. They found a way to turn us *all* into computer kids by lowering the cost and increasing the value of entry to the point where microcomputers today affect all of our lives. And in doing so, they created an industry unlike any other.

This book is about that industry. It is not a history of the personal computer but rather all the parts of a history needed to understand how the industry functions, to put it in some context from which knowledge can be drawn. My job is to explain how this little part of the world really works. Historians have a harder job because they can be faulted for what is left out; explainers like me can get away with printing only the juicy parts.

Juice is my business. I write a weekly gossip column in *InfoWorld*, a personal computer newspaper. Think for a moment about what a bizarre concept that is—an industrial gossip column. Rumors and gossip become institutionalized in cultures that are in constant flux. Politics, financial markets, the entertainment industry, and the personal computer business live by rumors. But for gossip to play a role in a culture, it must both serve a useful function and have an audience that sees value in participation—in originating or spreading the rumor. Readers must feel they have a personal connection—whether it is to a stock price, Madonna's marital situation, or the impending introduction of a new personal computer.

And who am I to sit in judgment this way on an entire industry?

I'm a failure, of course.

It takes a failure—someone who is not quite clever enough to succeed or to be considered a threat—to gain access to the heart of any competitive, ego-driven industry. This is a business that won't brook rivals but absolutely demands an audience. I am that audience. I can program (poorly) in four computer languages, though all the computer world seems to care about anymore is a language called C. I have made hardware devices that almost worked. I qualify as the ideal informed audience for all those fragile geniuses who want their greatness to be understood and acknowledged.

About thirty times a week, the second phone on my desk rings. At the other end of that line, or at the sending station of an electronic mail message, or sometimes even on the stamp-licking end of a letter sent through the U.S. mail is a type of person literally unknown outside America. He—for the callers are nearly always male—is an engineer or programmer from a personal computer manufacturer or a software publisher. His purpose in calling is to share with me and with my 500,000 weekly readers the confidential product plans, successes, and failures of his company. Specifications, diagrams, parts lists, performance benchmarks—even computer programs—arrive regularly, invariably at the risk of somebody's job. One day it's a disgruntled Apple Computer old-timer, calling to bitch about the current management and by-the-way reveal the company's product plans for the next year. The next day it's a programmer from IBM's lab in Austin, Texas, calling to complain about an internal rivalry with another IBM lab in England and in the process telling all sorts of confidential information.

What's at work here is the principle that companies lie, bosses lie, but engineers are generally incapable of lying. If they lied, how could the many complex parts of a computer or a software application be expected to actually work together?

"Yeah, I know I said wire Y-21 would be 12 volts DC, but, heck, I lied."

Nope, it wouldn't work.

Most engineers won't even tolerate it when others in their companies lie, which is why I get so many calls from embarrassed or enraged techies undertaking what they view as damage control but their companies probably see as sabotage.

The smartest companies, of course, hide their engineers, never bringing them out in public, because engineers are not to be trusted:

Me: "Great computer! But is there any part of it you'd do differently if you could do it over again?"

Engineer: "Yup, the power supply. Put your hand on it right here. Feel how hot that is? Damn thing's so overloaded I'm surprised they haven't been bursting into flames all over the country.

I've got a fire extinguisher under the table just in case. Oh, I told the company about it, too, but would they listen?"

I love engineers.

This sort of thing doesn't happen in most other U.S. industries, and it never happens in Asia. Chemists don't call up the offices of *Plastics Design Forum* to boast about their new, top-secret thermoplastic alloy. The *Detroit Free Press* doesn't hear from engineers at Chrysler, telling about the bore and stroke of a new engine or in what car models that engine is likely to appear, and when. But that's exactly what happens in the personal computer industry.

Most callers fall into one of three groups. Some are proud of their work but are afraid that the software program or computer system they have designed will be mismarketed or never marketed at all. Others are ashamed of a bad product they have been associated with and want to warn potential purchasers. And a final group talks out of pure defiance of authority.

All three groups share a common feeling of efficacy: They believe that something can be accomplished by sharing privileged information with the world of microcomputing through me. What they invariably want to accomplish is a change in their company's course, pushing forward the product that might have been ignored, pulling back the one that was released too soon, or just showing management that it can be defied. In a smokestack industry, this would be like a couple of junior engineers at Ford taking it on themselves to go public with their conviction that next year's Mustang really ought to have fuel injection.

That's not the way change is accomplished at Ford, of course, where the business of business is taken very seriously, change takes place very slowly, and words like *ought* don't have a place outside the executive suite, and maybe not even there. Nor is change accomplished this way in the mainframe computer business, which moves at a pace that is glacial, even in comparison to

Ford. But in the personal computer industry, where few executives have traditional business backgrounds or training and a totally new generation of products is introduced every eighteen months, workers can become more committed to their creation than to the organization for which they work.

Outwardly, this lack of organizational loyalty looks bad, but it turns out to be very good. Bad products die early in the marketplace or never appear. Good products are recognized earlier. Change accelerates. And organizations are forced to be more honest. Most especially, everyone involved shares the same understanding of why they are working: to create the product.

The founders of the microcomputer industry were groups of boys who banded together to give themselves power. For the most part, they came from middle-class and upper-middle-class homes in upscale West Coast communities. They weren't rebels; they resented their parents and society very little. Their only alienation was the usual hassle of the adolescent—a feeling of being prodded into adulthood on somebody else's terms. So they split off and started their own culture, based on the completely artificial but totally understandable rules of computer architecture. They defined, built, and controlled (and still control) an entire universe in a box—an electronic universe of ideas rather than people—where they made all the rules, and could at last be comfortable. They didn't resent the older people around them—you and me, the would-be customers—but came to pity us because we couldn't understand the new order inside the box—the microcomputer.

And turning this culture into a business? That was just a happy accident that allowed these boys to put off forever the hor-

ror age—that dividing line to adulthood that they would otherwise have been forced to cross after college.

The 1980s were not kind to America. Sitting at the end of the
longest period of economic expansion in history, what have we
gained? Budget deficits are bigger. Trade deficits are bigger. What
property we haven't sold we've mortgaged. Our basic industries
are being moved overseas at an alarming rate. We pretended for a
time that junk bond traders and corporate disassemblers create
wealth, but they don't. America is turning into a service economy and telling itself that's good. But it isn't.

America was built on the concept of the frontier. We carved
a nation out of the wilderness, using as tools enthusiasm, adolescent energy, and an unwillingness to recognize limitations. But
we are running out of recognized frontiers. We are getting older
and stodgier and losing our historic advantage in the process. In
contrast, the PC business is its own frontier, created inside the
box by inward-looking nerds who could find no acceptable challenge in the adult world. Like any other true pioneers, they don't
care about what is possible or not possible; they are dissatisfied
with the present and excited about the future. They are anti-
establishment and rightly see this as a prerequisite for success.

Time after time, Japanese companies have aimed at dominating the PC industry in the same way that similar programs
have led to Japanese success in automobiles, steel, and consumer
electronics. After all, what is a personal computer but a more
expensive television, calculator, or VCR? With the recent exception of laptop computers, though, Japan's luck has been poor in
the PC business. Korea, Taiwan, and Singapore have fared similarly and are still mainly sources of cheap commodity components that go into American-designed and -built PCs.

As for the Europeans, they are obsessed with style, thinking
that the external design of a computer is as important as its raw

performance. They are wrong: horsepower sells. The results are high-tech toys that look pretty, cost a lot, and have such low performance that they suggest Europe hasn't quite figured out what PCs are even used for.

It's not that the Japanese and others can't build personal computers as well as we can; manufacturing is what they do best. What puts foreigners at such a disadvantage is that they usually don't know what to build because the market is changing so quickly; a new generation of machines and software appears every eighteen months.

The Japanese have grown rich in other industries by moving into established markets with products that are a little better and a little cheaper, but in the PC business the continual question that needs asking is, "Better than what?" Last year's model? this year's? next year's? By the time the Asian manufacturers think they have a sense of what to aim for, the state of the art has usually changed.

In the PC business, constant change is the only norm, and adolescent energy is the source of that change.

The Japanese can't take over because they are too grown-up. They are too businesslike, too deliberate, too slow. They keep trying, with little success, to find some level at which it all makes sense. But that level does not exist in this business, which has grown primarily without adult supervision.

Smokestacks, skyscrapers, half-acre mahogany desks, corporate jets, gray hair, the building of things in enormous factories by crowds of faceless, time card–punching workers: these are traditional images of corporate success, even at old-line computer companies like IBM.

Volleyball, junk food, hundred-hour weeks, cubicles instead of offices, T-shirts, factories that either have no workers or run, unseen, in Asia: these are images of corporate success in the personal computer industry today.

The differences in corporate culture are so profound that IBM has as much in common with Tehran or with one of the newly discovered moons of Neptune as it does with a typical personal computer software company. On August 25, 1989, for example, all 280 employees of Adobe Systems Inc., a personal computer software company, armed themselves with waste baskets and garden hoses for a company-wide water fight to celebrate the shipping of a new product. Water fights don't happen at General Motors, Citicorp, or IBM, but then those companies don't have Adobe's gross profit margins of 43 percent either.

We got from boardrooms to water balloons led not by a Tom Watson, a Bill Hewlett, or even a Ross Perot but by a motley group of hobbyist/opportunists who saw a niche that needed to be filled. Mainly academics and nerds, they had no idea how businesses were supposed to be run, no sense of what was impossible, so they faked it, making their own ways of doing business—ways that are institutionalized today but not generally documented or formally taught. It's the triumph of the nerds.

Here's the important part: they are *our* nerds. And having, by their conspicuous success, helped create this mess we're in, they had better have a lot to teach us about how to recreate the business spirit we seem to have lost.

THE TYRANNY OF THE NORMAL DISTRIBUTION

This chapter is about smart people. My own, highly personal definition of what it means to be smart has changed over the years. When I was in the second grade, smart meant being able to read a word like *Mississippi* and then correctly announce how many syllables it had (four, right?). During my college days, smart people were the ones who wrote the most complex and amazing computer programs. Today, at college plus twenty years or so, my definition of smart means being able to deal honestly with people yet somehow avoid the twin perils of either pissing them off or of committing myself to a lifetime of indentured servitude by trying too hard to be nice. In all three cases, being smart means accomplishing something beyond my current level of ability, which is probably the way most other folks define it. Even you.

But what if *nothing* is beyond your ability? What if you've got so much brain power that little things like getting through school and doing brain surgery (or getting through school *while* doing brain surgery) are no big sweat? Against what, then, do you measure yourself?

Back in the 1960s at MIT, there was a guy named Harvey Allen, a child of privilege for whom everything was just that easy, or at least that's the way it looked to his fraternity brothers. Every Sunday morning, Harvey would wander down to the frat house dining room and do the *New York Times* crossword puzzle before breakfast—the *whole* puzzle, even to the point of knowing off the top of his head that Nunivak is the seven-letter name for an island in the Bering Sea off the southwestern coast of Alaska.

One of Harvey Allen's frat brothers was Bob Metcalfe, who noticed this trick of doing crossword puzzles in the time it took the bacon to fry and was in awe. Metcalfe, no slouch himself, eventually received a Ph.D., invented the most popular way of linking computers together, started his own company, became a multimillionaire, put his money and name on two MIT professorships, moved into a 10,000-square-foot Bernard Maybeck mansion in California, and *still* can't finish the *New York Times* crossword, which continues to be his definition of pure intelligence.

Not surprisingly, Harvey Allen hasn't done nearly as much with his professional life as Bob Metcalfe has because Harvey Allen had less to prove. After all, he'd already done the crossword puzzle.

Now we're sitting with Matt Ocko, a clever young programmer who is working on the problem of seamless communication between programs running on all different types of computers, which is something along the lines of getting vegetables to talk with each other even when they don't want to. It's a big job, but Matt says he's just the man to do it.

Back in North Carolina, Matt started DaVinci Systems to produce electronic mail software. Then he spent a year working as a programmer at Microsoft. Returning to DaVinci, he wrote an electronic mail program now used by more than 500,000 people, giving Matt a net worth of $1.5 million. Eventually he joined a new company, UserLand Software, to work on the problem of teaching vegetables to talk. And somewhere in there, Matt Ocko went to Yale. He is 22 years old.

Sitting in a restaurant, Matt drops every industry name he can think of and claims at least tangential involvement with every major computer advance since before he was born. Synapses snapping, neurons straining near the breaking point—for some reason he's putting a terrific effort into making me believe what I always knew to be true: Matt Ocko is a smart kid. Like Bill Gates, he's got something to prove. I ask him if he ever does the *New York Times* crossword.

Personal computer hardware and software companies, at least the ones that are doing new and interesting work, are all built around technical people of extraordinary ability. They are a mixture of Harvey Allens and Bob Metcalfes—people who find creativity so effortless that invention becomes like breathing or who have something to prove to the world. There are more Bob Metcalfes in this business than Harvey Allens but still not enough of either type.

Both types are exceptional. They are the people who are left unchallenged by the simple routine of making a living and surviving in the world and are capable, instead, of first imagining and then making a living from whole new worlds they've created in the computer. When balancing your checking account isn't, by itself, enough, why not create an alternate universe where checks don't exist, nobody really dies, and monsters can be killed by jumping on their heads? That's what computer game designers do. They define what it means to be a sky and a wall and a man, and to have color, and what should happen when man and monster collide, while the rest of us just try to figure out whether interest rates have changed enough to justify refinancing our mortgages.

Who are these ultrasmart people? We call them engineers, programmers, hackers, and techies, but mainly we call them nerds.

Here's your father's image of the computer nerd: male, a sloppy dresser, often overweight, hairy, and with poor interpersonal communication skills. Once again, Dad's wrong. Those who work with nerds but who aren't themselves programmers or engineers imagine that nerds are withdrawn—that is, until they have some information the nerd needs or find themselves losing an argument with him. Then they learn just how expressive a nerd can be. Nerds are expressive and precise in the extreme but only when they feel like it. They look the way they do as a deliberate statement about personal priorities, not because they're lazy. Their mode of communication is so precise that they can seem almost unable to communicate. Call a nerd Mike when he calls himself Michael and he likely won't answer, since you couldn't possibly be referring to him.

Out on the grass beside the Department of Computer Science at Stanford University, a group of computer types has been meeting every lunchtime for years and years just to juggle together. Groups of two, four, and six techies stand barefoot in the grass, surrounded by Rodin sculptures, madly flipping Indian clubs through the air, apparently aiming at each other's heads. As a spectator, the big thrill is to stand in the middle of one of these unstable geometric forms, with the clubs zipping past your head, experiencing what it must be like to be the nucleus of an especially busy atom. Standing with your head in their hands is a good time, too, to remember that these folks are not the way they look. They are precise, careful, and . . .

POW!!

"Oh, SHIT!!!!!!"

"Sorry, man. You okay?"

One day in the mid-1980s, *Time, Newsweek,* and the *Wall Street Journal* simultaneously discovered the computer culture, which they branded instantly and forever as a homogenized group they called nerds, who were supposed to be uniformly dressed in T-shirts and reeking of Snickers bars and Jolt cola.

Or just reeking. Nat Goldhaber, who founded a software company called TOPS, used to man his company's booth at computer trade shows. Whenever a particularly foul-smelling man would come in the booth, Goldhaber would say, "You're a programmer, aren't you?" "Why, yes," he'd reply, beaming at being recognized as a stinking god among men.

The truth is that there are big differences in techie types. The hardware people are radically different from the software people, and on the software side alone, there are at least three subspecies of programmers, two of which we are interested in here.

Forget about the first subspecies, the *lumpenprogrammers*, who typically spend their careers maintaining mainframe computer code at insurance companies. Lumpenprogrammers don't even like to program but have discovered that by the simple technique of leaving out the comments—clues, labels, and directions written in English—they are supposed to sprinkle in among their lines of computer code, their programs are rendered undecipherable by others, guaranteeing them a lifetime of dull employment.

The two programmer subspecies that *are* worthy of note are the hippies and the nerds. Nearly all great programmers are one type or the other. Hippie programmers have long hair and deliberately, even pridefully, ignore the seasons in their choice of clothing. They wear shorts and sandals in the winter and T-shirts all the time. Nerds are neat little anal-retentive men with penchants for short-sleeved shirts and pocket protectors. Nerds carry calculators; hippies borrow calculators. Nerds use decongestant nasal sprays; hippies snort cocaine. Nerds typically know forty-six different ways to make love but don't know any women.

Hippies know women.

In the actual doing of that voodoo that they do so well, there's a major difference, too, in the way that hippies and nerds write computer programs. Hippies tend to do the right things poorly; nerds tend to do the wrong things well. Hippie programmers are very good at getting a sense of the correct shape of a problem and how to solve it, but when it comes to the actual code writing, they can get sloppy and make major errors through pure boredom. For hippie programmers, the problem is solved when they've figured out *how* to solve it rather than later, when the work is finished and the problem no longer exists. Hippies live in a world of ideas. In contrast, the nerds are so tightly focused on the niggly details of making a program feature work efficiently that they can completely fail to notice major flaws in the overall concept of the project.

Conventional wisdom says that asking hippies and nerds to work together might lead to doing the wrong things poorly, but that's not so. With the hippies dreaming and the nerds coding, a good combination of the two can help keep a software development project both on course and on schedule. The real problem is finding such superprogrammers in the first place. Often they hide.

Back in the 1950s, a Harvard psychologist named George A. Miller wrote "The Magical Number Seven, Plus or Minus Two," a landmark journal article. Miller studied short-term memory, especially the quick memorization of random sequences of numbers. He wanted to know, going into the study, how many numbers people could be reliably expected to remember a few minutes after having been told those numbers only once.

The answer—the magical number—was about seven. Grab

some people off the street, tell them to remember the numbers 2-4-3-5-1-8-3 in that order, and most of them could, at least for a while. There was variation in ability among Miller's subjects, with some people able to remember eight or nine numbers and an equal number of people able to remember only five or six numbers, so he figured that seven (plus or minus two) numbers accurately represented the ability range of nearly the entire population.

Miller's concept went beyond numbers, though, to other organizations of data. For example, most of us can remember about seven recently learned pieces of similarly classified data, like names, numbers, or clues in a parlor game.

You're exposed to Miller's work every time you dial a telephone, because it was a factor in AT&T's decision to standardize on seven-digit local telephone numbers. Using longer numbers would have eliminated the need for area codes, but then no one would ever be able to remember a number without first writing it down.

Even area codes follow another bit of Miller's work. He found that people could remember more short-term information if they first subdivided the information into pieces—what Miller called "chunks." If I tell you that my telephone number is (415) 525-9270 (it is; call any time), you probably remember the area code as a separate chunk of information, a single data point that doesn't significantly affect your ability to remember the seven-digit number that follows. The area code is stored in memory as a single three-digit number—415—related to your knowledge of geography and the telephone system that rather than the random sequence of one-digit numbers—4-1-5—that relate to nothing in particular.

We store and recall memories based on their content, which explains why jokes are remembered by their punch lines, eliminating the possibility of mistaking "Why did the chicken cross the road?'' with ''How do you get to Carnegie Hall?'' It's also why remembering your way home doesn't interfere with remembering your way to the bathroom: the sets of information are maintained as different chunks in memory.

Some very good chess players use a form of chunking to keep track of the progress of a game by taking it to a higher level of abstraction in their minds. Instead of remembering the changing positions of each piece on the board, they see the game in terms of flowing trends, rather like the intuitive grammar rules that most of us apply without having to know their underlying definitions. But the very best chess players don't play this way at all: they effortlessly remember the positions of all the pieces.

As in most other statistical studies, Miller used a random sample of a few hundred subjects intended to represent the total population of the world. It was cheaper than canvassing the whole planet, and not significantly less accurate. The study relied on Miller's assurance that the population of the sample studied and that of the world it represented were both ''normal''—a statistical term that allows us to generalize accurately from a small, random sample to a much larger population from which that sample has been drawn.

Avoiding a lengthy explanation of bell-shaped curves and standard deviations, please trust George Miller and me when we tell you that this means 99.7 percent of all people can remember seven (plus or minus two) numbers. Of course, that leaves 0.3 percent, or 3 out of every 1,000 people, who can remember either fewer than five numbers or more than nine. As true believers in the normal distribution, we know it's symmetrical, which means that just about as many people can remember more than nine numbers as can remember fewer than five.

In fact, there are learning-impaired people who can't remember even one number, so it should be no surprise that 0.15 percent, or 3 out of every 2,000 people, can remember fewer than five numbers, given Miller's test. Believe me, those three people are not likely to be working as computer programmers.

It is the 0.15 percent on the other side of the bell curve that we're interested in—the 3 out of every 2,000 people who can remember more than nine numbers. There are approximately 375,000 such people living in the United States, and most of them would make *terrific* computer programmers, if only we could find them.

So here's my plan for leading the United States back to dominance of the technical world. We'll run a short-term memory contest. I like the idea of doing it like those correspondence art schools that advertise on matchbook covers and run ads in women's magazines and *Popular Mechanics*—you know, the ones that want you to "draw Skippy."

"Win Big Bucks Just by Remembering 12 Numbers!" our matchbooks would say.

Wait, I have a better idea! We could have the contest live on national TV, and the viewers would call in on a 900 number that would cost them a couple of bucks each to play. We'd find thousands of potential top programmers who all this time were masquerading as truck drivers and cotton gin operators and beauticians in Cheyenne, Wyoming—people you'd never in a million years know were born to write software. The program would be self-supporting, too, since we know that less than 1 percent of the players would be winners. And the best part of all about this plan is that it's my idea. I'll be *rich!*

Behind my dreams of glory lies the fact that nearly all of the best computer programmers and hardware designers are people who would fall off the right side of George Miller's bell curve of short-term memory ability. This doesn't mean that being able to remember more than nine numbers at a time is a prerequisite for writing a computer program, just that being able to remember more than nine numbers at a time is probably a prerequisite for writing a really *good* computer program.

Writing software or designing computer hardware requires keeping track of the complex flow of data through a program or a machine, so being able to keep more data in memory at a time can be very useful. In this case, the memory we are talking about is the programmer's, not the computer's.

The best programmers find it easy to remember complex things. Charles Simonyi, one of the world's truly great programmers, once lamented the effect age was having on his ability to remember. "I have to really concentrate, and I might even get a headache just trying to imagine something clearly and distinctly with twenty or thirty components," Simonyi said. "When I was young, I could easily imagine a castle with twenty rooms with each room having ten different objects in it. I can't do that anymore."

Stop for a moment and look back at that last paragraph. George Miller showed us that only 3 in 2,000 people can remember more than nine simultaneous pieces of short-term data, yet Simonyi looked wistfully back at a time when he could remember 200 pieces of data, and still claimed to be able to think simultaneously of 30 distinct data points. Even in his doddering middle age (Simonyi is still in his forties), that puts the Hungarian so far over on the right side of Miller's memory distribution that he is barely on the same planet with the rest of us. And there are better programmers than Charles Simonyi.

Here is a fact that will shock people who are unaware of the way computers and software are designed: at the extreme edges of the normal distribution, there are programmers who are *100 times more productive* than the average programmer simply on the basis of the number of lines of computer code they can write in a given period of time. Going a bit further, since some programmers are so accomplished that their programming feats are beyond the ability of most of their peers, we might say that they are infinitely more productive for really creative, leading-edge projects.

The trick to developing a new computer or program, then, is not to hire a lot of smart people *but to hire a few very smart people.* This rule lies at the heart of most successful ventures in the personal computer industry.

Programs are written in a code that's referred to as a computer language, and that's just what it is—a language, complete with subjects and verbs and all the other parts of speech we used to be able to name back in junior high school. Programmers learn to speak the language, and good programmers learn to speak it fluently. The very best programmers go beyond fluency to the level of art, where, like Shakespeare, they create works that have value beyond that even recognized or intended by the writer. Who will say that Shakespeare isn't worth a dozen lesser writers, or a hundred, or a thousand? And who can train a Shakespeare? Nobody; they have to be born.

But in the computer world, there can be such a thing as having too much gray matter. Most of us, for example, would decide that Bob Metcalfe was more successful in his career than Harvey Allen, but that's because Metcalfe had things to prove to himself and the world, while Harvey Allen, already supreme, did not.

Metcalfe chose being smart as his method of gaining revenge against those kids who didn't pick him for their athletic teams back in school on Long Island, and he used being smart as a weapon against the girls who broke his heart or even in retaliation for the easy grace of Harvey Allen. Revenge is a common motivation for nerds who have something to prove.

The Harvey Allens of the world can apply their big brains to self-delusion, too, with great success. Donald Knuth is a Stanford computer science professor generally acknowledged as having the biggest brain of all—so big that it is capable on occasion of seeing things that aren't really there. Knuth, a nice guy whose first-ever publication was "The Potrszebie System of Weights and Measures" ("one-millionth of a potrszebie is a farshimmelt potrszebie"), in the June 1957 issue of *Mad* magazine, is better known for his multivolume work *The Art of Computer Programming*, the seminal scholarly work in his field.

The first volume of Knuth's series (dedicated to the IBM 650 computer, "in remembrance of many pleasant evenings") was printed in the late 1960s using old-fashioned but beautiful hot-type printing technology, complete with Linotype machines and the sharp smell of molten lead. Volume 2, which appeared a few years later, used photo-offset printing to save money for the publisher (the publisher of this book, in fact). Knuth didn't like the change from hot type to cold, from Lino to photo, and so he took a few months off from his other work, rolled up his sleeves, and set to work computerizing the business of setting type and designing type fonts. Nine years later, he was done.

Knuth's idea was that through the use of computers, photo offset, and especially the printing of numbers and mathematical formulas, could be made as beautiful as hot type. This was like Perseus giving fire to humans, and as ambitious, though well within the capability of Knuth's largest of all brains.

He invented a text formatting language called T_eX, which could drive a laser printer to place type images on the page as well as or better than the old linotype, and he invented another language, Metafont, for designing whole families of fonts. Draw a letter "A," and Metafont could generate a matching set of the other twenty-five letters of the alphabet.

When he was finished, Don Knuth saw that what he had done was good, and said as much in volume 3 of *The Art of Computer Programming*, which was typeset using the new technology.

It was a major advance, and in the introduction he proudly claimed that the printing once again looked just as good as the hot type of volume 1.

Except it didn't.

Reading his introduction to volume 3, I had the feeling that Knuth was wearing the emperor's new clothes. Squinting closely at the type in volume 3, I saw the letters had that telltale look of a low-resolution laser printer—not the beautiful, smooth curves of real type or even of a photo typesetter. There were "jaggies"—little bumps that make all the difference between good type and bad. Yet here was Knuth, writing the same letters that I was reading, and claiming that they were beautiful.

"Donnie," I wanted to say. "What are you talking about? Can't you see the jaggies?"

But he couldn't. Donald Knuth's gray matter, far more powerful than mine, was making him look beyond the actual letters and words to the mathematical concepts that underlay them. Had a good enough laser printer been available, the printing would have been beautiful, so that's what Knuth saw and I didn't. This effect of mind over what matters is both a strength and a weakness for those, like Knuth, who would break radical new ground with computers.

Unfortunately for printers, most of the rest of the world sees like me. The tyranny of the normal distribution is that we run the world as though it was populated entirely by Bob Cringelys, completely ignoring the Don Knuths among us. Americans tend to look at research like George Miller's and use it to custom-design cultural institutions that work at our most common level of mediocrity—in this case, the number seven. We cry about Japanese or Korean students, having higher average math scores in high school than do American students. "Oh, no!" the editorials scream. "Johnny will never learn FORTRAN!" In fact, average high school math scores have little bearing on the state of basic research or of product research and development in Japan, Korea, or the United States. What really matters is what we do with the edges of the distribution rather than the middle. Whether Johnny learns FORTRAN is relevant only to Johnny, not to America. Whether Johnny learns to *read* matters to America.

This mistaken trend of attributing average levels of competence or commitment to the whole population extends far beyond human memory and computer technology to areas like medicine. Medical doctors, for example, say that spot weight reduction is not possible. "You can reduce body fat overall through dieting and exercise, but you can't take fat just off your butt," they lecture. Bodybuilders, who don't know what the doctors know, have been doing spot weight reduction for years. What the doctors don't say out loud when they make their pronouncements on spot reduction is that their definition of exercise is 20 minutes, three times a week. The bodybuilder's definition of exercise is more like 5 to 7 hours, five times a week —up to thirty-five times as much.

Doctors might protest that average people are unlikely to spend 35 hours per week exercising, but that is exactly the point: Most of us wouldn't work 36 straight hours on a computer program either, but there are programmers and engineers who thrive on working that way.

Average populations will always achieve only average results, but what we are talking about are exceptional populations seeking extraordinary results. In order to make spectacular progress, to achieve profound results in nearly any field, what is required is a combination of unusual ability and profound dedication—very unaverage qualities for a population that typically spends 35 hours per week watching television and less than 1 hour exercising.

Brilliant programmers and champion bodybuilders already

have these levels of ability and motivation in their chosen fields. And given that we live in a society that can't seem to come up with coherent education or exercise policies, it's good that the hackers and iron-pumpers are self-motivated. Hackers will seek out and find computing problems that challenge them. Bodybuilders will find gyms or found them. We don't have to change national policy to encourage bodybuilders or super-programmers.

All we have to do is stay out of their way.

▸   ▸   ▸   ▸   ▸   ▸   ▸   ▸   ▸   ▸   ▸   ▸

WHY THEY DON'T CALL
IT COMPUTER VALLEY

Reminders of just how long I've been around this youth-driven business keep hitting me in the face. Not long ago I was poking around a store called the Weird Stuff Warehouse, a sort of Silicon Valley thrift shop where you can buy used computers and other neat junk. It's right across the street from Fry's Electronics, the legendary computer store that fulfills every need of its techie customers by offering rows of junk food, soft drinks, girlie magazines, and Maalox, in addition to an enormous selection of new computers and software. You can't miss Fry's; the building is painted to look like a block-long computer chip. The front doors are labeled *Enter* and *Escape*, just like keys on a computer keyboard.

Weird Stuff, on the other side of the street, isn't painted to look like anything in particular. It's just a big storefront filled with tables and bins holding the technological history of Silicon Valley. Men poke through the ever-changing inventory of junk while women wait near the door, rolling their eyes and telling each other stories about what stupid chunk of hardware was dragged home the week before.

Next to me, a gray-haired member of the short-sleeved sport shirt and Hush Puppies school of 1960s computer engineering was struggling to drag an old printer out from under a table so he could show his 8-year-old grandson the connector he'd designed a lifetime ago. Imagine having as your contribution to history the fact that pin 11 is connected to a red wire, pin 18 to a blue wire, and pin 24 to a black wire.

On my own search for connectedness with the universe, I came across a shelf of Apple III computers for sale for $100 each. Back in 1979, when the Apple III was still six months away from being introduced as a $3,000 office computer, I remember sitting in a movie theater in Palo Alto with one of the Apple III designers, pumping him for information about it.

There were only 90,000 Apple III computers ever made, which sounds like a lot but isn't. The Apple III had many problems, including the fact that the automated machinery that inserted dozens of computer chips on the main circuit board didn't push them into their sockets firmly enough. Apple's answer was to tell 90,000 customers to pick up their Apple III carefully, hold it twelve to eighteen inches above a level surface, and then drop it, hoping that the resulting crash would reseat all the chips.

Back at the movies, long before the Apple III's problems, or even its potential, were known publicly, I was just trying to get my friend to give me a basic description of the computer and its software. The film was *Barbarella*, and all I can remember now about the movie or what was said about the computer is this image of Jane Fonda floating across the screen in simulated weightlessness, wearing a costume with a clear plastic midriff. But then the rest of the world doesn't remember the Apple III at all.

It's this relentless throwing away of old technology, like the nearly forgotten Apple III, that characterizes the personal computer business and differentiates it from the business of building big computers, called mainframes, and minicomputers. Mainframe technology lasts typically twenty years; PC technology dies and is reborn every eighteen months.

There were computers in the world long before we called any of them "personal." In fact, the computers that touched our lives before the mid-1970s were as impersonal as hell. They sat in big air-conditioned rooms at insurance companies, phone companies, and the IRS, and their main function was to screw up our lives by getting us confused with some other guy named Cringely, who was a deadbeat, had a criminal record, and didn't much like to pay parking tickets. Computers were instruments of government and big business, and except for the punched cards that came in the mail with the gas bill, which we were supposed to return obediently with the money but without any folds, spindling, or mutilation, they had no physical presence in our lives.

How did we get from big computers that lived in the basement of office buildings to the little computers that live on our desks today? We didn't. Personal computers have almost nothing to do with big computers. They never have.

A personal computer is an electronic gizmo that is built in a factory and then sold by a dealer to an individual or a business. If everything goes as planned, the customer will be happy with the purchase, and the company that makes the personal computer, say Apple or Compaq, won't hear from that customer again until he or she buys another computer. Contrast that with the mainframe computer business, where big computers are built in a factory, sold directly to a business or government, installed by the computer maker, serviced by the computer maker (for a monthly fee), financed by the computer maker, and often running software written by the computer maker (and licensed, not sold, for another monthly fee). The big computer company makes as much money from servicing, financing, and programming the computer as it does from selling it. It not only wants to continue to know the customer, it wants to be in the customer's dreams.

The only common element in these two scenarios is the factory. Everything else is different. The model for selling personal computers is based on the idea that there are millions of little customers out there; the model for selling big computers has always been based on the idea that there are only a few large customers.

When IBM engineers designed the System 650 mainframe in the early 1950s, their expectation was to build fifty in all, and the cost structure that was built in from the start allowed the company to make a profit on only fifty machines. Of course, when computers became an important part of corporate life, IBM found itself selling far more than fifty—1,500, in fact—with distinct advantages of scale that brought gross profit margins up to the 60 to 70 percent range, a range that computer companies eventually came to expect. So why bother with personal computers?

Big computers and little computers are completely different beasts created by radically different groups of people. It's logical, I know, to assume that the personal computer came from shrinking a mainframe, but that's not the way it happened. The PC business actually grew up from the semiconductor industry. Instead of being a little mainframe, the PC is, in fact, more like an incredibly big chip. Remember, they don't call it Computer Valley. They call it Silicon Valley, and it's a place that was invented one afternoon in 1957 when Bob Noyce and seven other engineers quit en masse from Shockley Semiconductor.

William Shockley was a local boy and amateur magician who had gone on to invent the transistor at Bell Labs in the late 1940s and by the mid-1950s was on his own building transistors in what had been apricot drying sheds in Mountain View, California.

Shockley was a good scientist but a bad manager. He posted a list of salaries on the bulletin board, pissing off those who were being paid less for the same work. When the work wasn't going well, he blamed sabotage and demanded lie detector tests. That did it. Just weeks after they'd toasted Shockley's winning the Nobel Prize in physics by drinking champagne over breakfast at Dinah's Shack, a red clapboard restaurant on El Camino Real, the "Traitorous Eight," as Dr. S. came to call them, hit the road.

For Shockley, it was pretty much downhill from there; today he's remembered more for his theories of racial superiority and for starting a sperm bank for geniuses in the 1970s than for the breakthrough semiconductor research he conducted in the 1940s and 1950s. (Of course, with several fluid ounces of Shockley semen still sitting on ice, we may not have heard the last of the doctor yet.)

Noyce and the others started Fairchild Semiconductor, the archetype for every Silicon Valley start-up that has followed. They got the money to start Fairchild from a young investment banker named Arthur Rock, who found venture capital for the firm. This is the pattern that has been followed ever since as groups of technical types split from their old companies, pick up venture capital to support their new idea, and move on to the next start-up. More than fifty new semiconductor companies eventually split off in this way from Fairchild alone.

At the heart of every start-up is an argument. A splinter group inside a successful company wants to abandon the current product line and bet the company on some radical new technology. The boss, usually the guy who invented the current technology, thinks this idea is crazy and says so, wishing the splinter group well on their new adventure. If he's smart, the old boss even helps his employees to leave by making a minority investment in their new company, just in case they are among the 5 percent of start-ups that are successful.

The appeal of the start-up has always been that it's a small operation, usually led by the smartest guy in the room but with the assistance of all players. The goals of the company are those of its people, who are all very technically oriented. The character of the company matches that of its founders, who were inevitably engineers—regular guys. Noyce was just a preacher's kid from Iowa, and his social sensibilities reflected that background.

There was no social hierarchy at Fairchild—no reserved parking spaces or executive dining rooms—and that remained true even later when the company employed thousands of workers and Noyce was long gone. There was no dress code. There were hardly any doors; Noyce had an office cubicle, built from shoulder-high partitions, just like everybody else. Thirty years later, he still had only a cubicle, along with limitless wealth.

They use cubicles, too, at Hewlett-Packard, which at one point in the late 1970s had more than 50,000 employees, but only three private offices. One office belonged to Bill Hewlett, one to David Packard, and the third to a guy named Paul Ely, who annoyed so many coworkers with his bellowing on the telephone that the company finally extended his cubicle walls clear to the ceiling. It looked like a freestanding elevator shaft in the middle of a vast open office.

The Valley is filled with stories of Bob Noyce as an Everyman with deep pockets. There was the time he stood in a long line at his branch bank and then asked the teller for a cashier's check for $1.3 million from his personal savings, confiding gleefully that he was going to buy a Learjet that afternoon. Then, after his divorce and remarriage, Noyce tried to join the snobbish Los Altos Country Club, only to be rejected because the club did not approve of his new wife, so he wrote another check and simply duplicated the country club facilities on his own property, within sight of the Los Altos clubhouse. "To hell with them," he said.

As a leader, Noyce was half high school science teacher and

half athletic team captain. Young engineers were encouraged to speak their minds, and they were given authority to buy whatever they needed to pursue their research. No idea was too crazy to be at least considered, because Noyce realized that great discoveries lay in crazy ideas and that rejecting out of hand the ideas of young engineers would just hasten that inevitable day when they would take off for their own start-up.

While Noyce's ideas about technical management sound all too enlightened to be part of anything called big business, they worked well at Fairchild and then at Noyce's next creation, Intel. Intel was started, in fact, because Noyce couldn't get Fairchild's eastern owners to accept the idea that stock options should be a part of compensation for all employees, not just for management. He wanted to tie everyone, from janitors to bosses, into the overall success of the company, and spreading the wealth around seemed the way to go.

This management style still sets the standard for every computer, software, and semiconductor company in the Valley today, where office doors are a rarity and secretaries hold shares in their company's stock. Some companies follow the model well, and some do it poorly, but every CEO still wants to think that the place is being run the way Bob Noyce would have run it.

The semiconductor business is different from the business of building big computers. It costs a lot to develop a new semiconductor part but not very much to manufacture it once the design is proved. This makes semiconductors a volume business, where the most profitable product lines are those manufactured in the greatest volume rather than those that can be sold in smaller quantities with higher profit margins. Volume is everything.

To build volume, Noyce cut all Fairchild components to a uniform price of one dollar, which was in some cases not much more than the cost of manufacturing them. Some of Noyce's partners thought he was crazy, but volume grew quickly, followed by profits, as Fairchild expanded production again and again to meet demand, continually cutting its cost of goods at the same time. The concept of continually dropping electronic component prices was born at Fairchild. The cost per transistor dropped by a factor of 10,000 over the next thirty years.

To avoid building a factory that was 10,000 times as big, Noyce came up with a way to give customers more for their money while keeping the product price point at about the same level as before. While the cost of semiconductors was ever falling, the cost of electronic subassemblies continued to increase with the inevitably rising price of labor. Noyce figured that even this trend could be defeated if several components could be built together on a single piece of silicon, eliminating much of the labor from electronic assembly. It was 1959, and Noyce called his idea an *integrated circuit*. "I was lazy," he said. "It just didn't make sense to have people soldering together these individual components when they could be built as a single part."

Jack Kilby at Texas Instruments had already built several discrete components on the same slice of germanium, including the first germanium resistors and capacitors, but Kilby's parts were connected together on the chip by tiny gold wires that had to be installed by hand. TI's integrated circuit could not be manufactured in volume.

The twist that Noyce added was to deposit a layer of insulating silicon oxide on the top surface of the chip—this was called the "planar process" that had been invented earlier at Fairchild —and then use a photographic process to print thin metal lines on top of the oxide, connecting the components together on the chip. These metal traces carried current in the same way that Jack Kilby's gold wires did, but they could be printed on in a single step rather than being installed one at a time by hand.

Using their new photolithography method, Noyce and his

boys put first two or three components on a single chip, then ten, then a hundred, then thousands. Today the same area of silicon that once held a single transistor can be populated with more than a million components, all too small to be seen.

Tracking the trend toward ever more complex circuits, Gordon Moore, who cofounded Intel with Noyce, came up with Moore's Law: the number of transistors that can be built on the same size piece of silicon will double every eighteen months. Moore's Law still holds true. Intel's memory chips from 1968 held 1,024 bits of data; the most common memory chips today hold a thousand times as much—1,024,000 bits—and cost about the same.

The integrated circuit—the IC—also led to a trend in the other direction—toward higher price points, made possible by ever more complex semiconductors that came to do the work of many discrete components. In 1971, Ted Hoff at Intel took this trend to its ultimate conclusion, inventing the microprocessor, a single chip that contained most of the logic elements used to make a computer. Here, for the first time, was a programmable device to which a clever engineer could add a few memory chips and a support chip or two and turn it into a real computer you could hold in your hands. There was no software for this new computer, of course—nothing that could actually be done with it —but the computer could be held in your hands or even sold over the counter, and that fact alone was enough to force a paradigm shift on Silicon Valley.

It was with the invention of the microprocessor that the rest of the world finally disappointed Silicon Valley. Until that point, the kids at Fairchild, Intel, and the hundred other chipmakers that now occupied the southern end of the San Francisco peninsula had been farmers, growing chips that were like wheat from which the military electronics contractors and the computer companies could bake their rolls, bagels, and loaves of bread—their computers and weapon control systems. But with their invention of the microprocessor, the Valley's growers were suddenly harvesting something that looked almost edible by itself. It was as though they had been supplying for years these expensive bakeries, only to undercut them all by inventing the Twinkie.

But the computer makers didn't want Intel's Twinkies. Microprocessors were the most expensive semiconductor devices ever made, but they were still too cheap to be used by the IBMs, the Digital Equipment Corporations, and the Control Data Corporations. These companies had made fortunes by convincing their customers that computers were complex, incredibly expensive devices built out of discrete components; building computers around microprocessors would destroy this carefully crafted concept. Microprocessor-based computers would be too cheap to build and would have to sell for too little money. Worse, their lower part counts would increase reliability, hurting the service income that was an important part of every computer company's bottom line in those days.

And the big computer companies just didn't have the vision needed to invent the personal computer. Here's a scene that happened in the early 1960s at IBM headquarters in Armonk, New York. IBM chairman Tom Watson, Jr., and president Al Williams were being briefed on the concept of computing with video display terminals and time-sharing, rather than with batches of punch cards. They didn't understand the idea. These were intelligent men, but they had a firmly fixed concept of what computing was supposed to be, and it didn't include video display terminals. The briefing started over a second time, and finally a light bulb went off in Al Williams's head. "So what you are talking about is data processing but not in the same room!" he exclaimed.

IBM played for a short time with a concept it called teleprocessing, which put a simple computer terminal on an executive's desk, connected by telephone line to a mainframe computer somewhere. The idea was that the Big Boss would be able to look into the bowels of the company and know instantly how many widgets were being produced in the Muncie plant. That was the idea, but what IBM discovered from this mid-1960s exercise was that American business executives didn't know how to type and didn't want to learn. They had secretaries to type for them. No data were gathered on what middle managers would do with such a terminal because it wasn't aimed at them. Nobody even guessed that there would be millions of M.B.A.s hitting the streets over the following twenty years, armed with the ability to type and with the quantitative skills to use such a computing tool and to do some real damage with it. But that was yet to come, so exit teleprocessing, because IBM marketers chose to believe that this test indicated that American business executives would never be interested.

In order to invent a particular type of computer, you have to want first to use it, and the leaders of America's computer companies did not want a computer on their desks. Watson and Williams *sold* computers but they didn't use them. Williams's specialty was finance; it was through his efforts that IBM had turned computer leasing into a goldmine. Watson was the son of God—Tom Watson Sr.—and had been bred to lead the blue-suited men of IBM, not to design or use computers. Watson and Williams didn't have computer terminals at their desks. They didn't even work for a company that *believed* in terminals. Their concept was of data processing, which at IBM meant piles of paper cards punched with hundreds of rectangular, not round, holes. Round holes belonged to Univac.

The computer companies for the most part rejected the microprocessor, calling it too simple to perform their complex mainframe voodoo. It was an error on their part, and not lost on the next group of semiconductor engineers who were getting ready to explode from their current companies into a whole new generation of start-ups. This time they built more than just chips and ICs; they built entire computers, still following the rules for success in the semiconductor business: continual product development; a new family of products every year or two; ever increasing functionality; ever decreasing price for the same level of function; standardization; and volume, volume, volume.

It takes society thirty years, more or less, to absorb a new information technology into daily life. It took about that long to turn movable type into books in the fifteenth century. Telephones were invented in the 1870s but did not change our lives until the 1900s. Motion pictures were born in the 1890s but became an important industry in the 1920s. Television, invented in the mid-1920s, took until the mid-1950s to bind us to our sofas.

We can date the birth of the personal computer somewhere between the invention of the microprocessor in 1971 and the introduction of the Altair hobbyist computer in 1975. Either date puts us today about halfway down the road to personal computers' being a part of most people's everyday lives, which should be consoling to those who can't understand what all the hullabaloo is about PCs. Don't worry; you'll understand it in a few years, by which time they'll no longer be called PCs.

By the time that understanding is reached, and personal computers have wormed into all our lives to an extent far greater than they are today, the whole concept of personal computing will probably have changed. That's the way it is with information technologies. It takes us quite a while to decide what to do with them.

Radio was invented with the original idea that it would replace telephones and give us wireless communication. That implies two-way communication, yet how many of us own radio transmitters? In fact, the popularization of radio came as a broadcast medium, with powerful transmitters sending the same message—entertainment—to thousands or millions of inexpensive radio receivers. Television was the same way, envisioned at first as a two-way visual communication medium. Early phonographs could record as well as play and were supposed to make recordings that would be sent through the mail, replacing written letters. The magnetic tape cassette was invented by Phillips for dictation machines, but we use it to hear music on Sony Walkmans. Telephones went the other direction, since Alexander Graham Bell first envisioned his invention being used to pipe music to remote groups of people.

The point is that all these technologies found their greatest success being used in ways other than were originally expected. That's what will happen with personal computers too. Fifteen years from now, we won't be able to function without some sort of machine with a microprocessor and memory inside. Though we probably won't call it a personal computer, that's what it will be.

It takes new ideas a long time to catch on—time that is mainly devoted to evolving the idea into something useful. This fact alone dumps most of the responsibility for early technical innovation in the laps of amateurs, who can afford to take the time. Only those who aren't trying to make money can afford to advance a technology that doesn't pay.

This explains why the personal computer was invented by hobbyists and supported by semiconductor companies, eager to find markets for their microprocessors, by disaffected mainframe programmers, who longed to leave their corporate/mainframe world and get closer to the machine they loved, and by a new class of counterculture entrepreneurs, who were looking for a way to enter the business world after years of fighting against it.

The microcomputer pioneers were driven primarily to create machines and programs for their own use or so they could demonstrate them to their friends. Since there wasn't a personal computer business as such, they had little expectation that their programming and design efforts would lead to making a lot of money. With a single strategic exception—Bill Gates of Microsoft—the idea of making money became popular only later.

These folks were pursuing adventure, not business. They were the computer equivalents of the barnstorming pilots who flew around America during the 1920s, putting on air shows and selling rides. Like the barnstormers had, the microcomputer pioneers finally discovered a way to live as they liked. Both the barnstormers and microcomputer enthusiasts were competitive and were always looking for something against which they could match themselves. They wanted independence and total control, and through the mastery of their respective machines, they found it.

Barnstorming was made possible by a supply of cheap surplus aircraft after World War I. Microcomputers were made possible by the invention of solid state memory and the microprocessor. Both barnstorming and microcomputing would not have happened without previous art. The barnstormers needed a war to train them and to leave behind a supply of aircraft, while microcomputers would not have appeared without mainframe computers to create a class of computer professionals and programming languages.

Like early pilots and motorists, the first personal computer drivers actually enjoyed the hazards of their primitive computing environments. Just getting from one place to another in an early automobile was a challenge, and so was getting a program to run on the first microcomputers. Breakdowns were frequent, even welcome, since they gave the enthusiast something to brag about to friends. The idea of doing real work with a microcomputer wasn't even considered.

Planes that were easy to fly, cars that were easy to drive, computers that were easy to program and use weren't nearly as interesting as those that were cantankerous. The test of the pioneer was how well he did despite his technology. In the computing arena, this meant that the best people were those who could most completely adapt to the idiosyncrasies of their computers. This explains the rise of arcane computer jargon and the disdain with which "real programmers" still often view computers and software that are easy to use. They interpret "ease of use" as "lack of challenge." The truth is that easy-to-use computers and programs take much more skill to produce than did the hairy-chested, primitive products of the mid-1970s.

Since there really wasn't much that could be done with microcomputers back then, the great challenge was found in overcoming the adversity involved in doing *anything*. Those who were able to get their computers and programs running at all went on to become the first developers of applications.

With few exceptions, early microcomputer software came from the need of some user to have software that did not yet exist. He needed it, so he invented it. And son of a gun, bragging about the program at his local computing club often dragged from the membership others who needed that software, too, wanted to buy it, and an industry was born.

AMATEUR HOUR

You have to wonder what it was we were doing before we had all these computers in our lives. Same stuff, pretty much. Down at the auto parts store, the counterman had to get a ladder and climb way the heck up to reach some top shelf, where he'd feel around in a little box and find out that the muffler clamps were all gone. Today he uses a computer, which tells him that there are three muffler clamps sitting in that same little box on the top shelf. But he still has to get the ladder and climb up to get them, and, worse still, sometimes the computer lies, and there are no muffler clamps at all, spoiling the digital perfection of the auto parts world as we have come to know it.

What we're often looking for when we add the extra overhead of building a computer into our businesses and our lives is certainty. We want something to believe in, something that will take from our shoulders the burden of knowing when to reorder muffler clamps. In the twelfth century, before there even were muffler clamps, such certainty came in the form of a belief in God, made tangible through the building of cathedrals—places where God could be accessed. For lots of us today, the belief is more in the sanctity of those digital zeros and ones, and our cathedral is the personal computer. In a way, we're replacing God with Bill Gates.

Uh-oh.

The problem, of course, is with those zeros and ones. Yes or no, right or wrong, is what those digital bits seem to signify, looking so clean and unconflicted that we forget for a moment about that time in the eighth grade when Miss Schwerko humiliated us all with a true-false test. The truth is, that for all the apparent precision of computers, and despite the fact that our mothers and Tom Peters would still like to believe that perfection is attainable in this life, computer and software companies are still remarkably imprecise places, and their products reflect it. And why shouldn't they, since we're still at the fumbling stage, where good and bad developments seem to happen at random.

Look at Intel, for example. Up to this point in the story, Intel comes off pretty much as high-tech heaven on earth. As the semiconductor company that most directly begat the personal computer business, Intel invented the microprocessor and memory technologies used in PCs and acted as an example of how a high-tech company should be organized and managed. But that doesn't mean that Bob Noyce's crew didn't screw up occasionally.

There was a time in the early 1980s when Intel suffered terrible quality problems. It was building microprocessors and other parts by the millions and by the millions these parts tested bad. The problem was caused by dust, the major enemy of computer chip makers. When your business relies on printing metallic traces that are only a millionth of an inch wide, having a dust mote ten times that size come rolling across a silicon wafer means that some traces won't be printed correctly and some parts won't work at all. A few bad parts are to be expected, since there are dozens, sometimes hundreds, printed on a single wafer, which is later cut into individual components. But Intel was suddenly getting as many bad parts as good, and that was bad for business.

Semiconductor companies fight dust by building their components in expensive clean rooms, where technicians wear surgical masks, paper booties, rubber gloves, and special suits and where the air is specially filtered. Intel had plenty of clean rooms, but it still had a big dust problem, so the engineers cleverly decided that the wafers were probably dusty *before* they ever arrived at Intel. The wafers were made in the East by Monsanto. Suddenly it was Monsanto's dust problem.

Monsanto engineers spent months and millions trying to eliminate every last speck of dust from their silicon wafer production facility in South Carolina. They made what they thought was terrific progress, too, though it didn't show in Intel's production yields, which were still terrible. The funny thing was that Monsanto's other customers weren't complaining. IBM, for example, wasn't complaining, and IBM was a very picky customer, always asking for wafers that were extra big or extra small or triangular instead of round. IBM was having no dust problems.

If Monsanto was clean and Intel was clean, the only remaining possibility was that the wafers somehow got dusty on their trip between the two companies, so the Monsanto engineers hired a private investigator to tail the next shipment of wafers to Intel. Their private eye uncovered an Intel shipping clerk who was opening incoming boxes of super-clean silicon wafers and then counting out the wafers by hand into piles on a super-unclean desktop, just to make sure that Bob Noyce was getting every silicon wafer he was paying for.

The point of this story goes far beyond the undeification of Intel to a fundamental characteristic of most high-tech businesses. There is a business axiom that management gurus spout and that bigshot industrialists repeat to themselves as a mantra if they want to sleep well at night. The axiom says that when a business grows past $1 billion in annual sales, it becomes too large for any one individual to have a significant impact. Alas, this is not true when it's a $1 billion high-tech business, where too often the critical path goes right through the head of one particular programmer or engineer or even through the head of a well-meaning clerk down in the shipping department. Remember that Intel was already a $1 billion company when it was brought to its knees by desk dust.

The reason that there are so many points at which a chip, a computer, or a program is dependent on just one person is that the companies lack depth. Like any other new industry, this is one staffed mainly by pioneers, who are, by definition, a small minority. People in critical positions in these organizations don't usually have backup, so when they make a mistake, the whole company makes a mistake.

My estimate, in fact, is that there are only about twenty-five real people in the entire personal computer industry—this shipping clerk at Intel and around twenty-four others. Sure, Apple Computer has 10,000 workers, or says it does, and IBM claims nearly 400,000 workers worldwide, but has to be lying. Those workers must be temps or maybe androids because I keep running into the same two dozen people at every company I visit. Maybe it's a tax dodge. Finish this book and you'll see; the companies keep changing, but the names are always the same.

Intel begat the microprocessor and the dynamic random access memory chip, which made possible MITS, the first of many personal computer companies with a stupid name. And MITS, in turn, made possible Microsoft, because computer hardware must exist, or at least be claimed to exist, before programmers can even envision software for it. Just as cave dwellers didn't squat with their flint tools chipping out parking brake assemblies for 1967 Buicks, so programmers don't write software that has no computer upon which to run. Hardware nearly always leads software, enabling new development, which is why Bill Gates's conversion from minicomputers to microcomputers did not come (could not come) until 1974, when he was a sophomore at Harvard University and the appearance of the MITS Altair 8800 computer made personal computer software finally possible.

Like the Buddha, Gates's enlightenment came in a flash. Walking across Harvard Yard while Paul Allen waved in his face the January 1975 issue of *Popular Electronics* announcing the Altair 8800 microcomputer from MITS, they both saw instantly that there would really be a personal computer industry and that the industry would need programming languages. Although there were no microcomputer software companies yet, 19-year-old Bill's first concern was that they were already too late. "We realized that the revolution might happen without us," Gates said. "After we saw that article, there was no question of where our life would focus."

"Our *life*?" What the heck does Gates mean here—that he and Paul Allen were joined at the frontal lobe, sharing a single life, a single set of experiences? In those days, the answer was "yes." Drawn together by the idea of starting a pioneering software company and each convinced that he couldn't succeed alone, they committed to sharing a single life—a life unlike that of most other PC pioneers because it was devoted as much to doing business as to doing technology.

Gates was a businessman from the start; otherwise, why would he have been worried about being passed by? There was plenty of room for high-level computer languages to be developed for the fledgling platforms, but there was only room for one *first* high-level language. Anyone could participate in a movement, but only those with the right timing could control it. Gates knew that the first language—the one resold by MITS, maker of the Altair—would become the standard for the whole industry. Those who seek to establish such de facto standards in any industry do so for business reasons.

"This is a very personal business, but success comes from appealing to groups," Gates says. "Money is made by setting de facto standards."

The Altair was not much of a consumer product. It came typically as an unassembled $350 kit, clearly targeting only the electronic hobbyist market. There was no software for the machine, so, while it may have existed, it sure didn't compute. There wasn't even a keyboard. The only way of programming the computer at first was through entering strings of hexadecimal code by flicking a row of switches on the front panel. There was no display other than some blinking lights. The Altair was limited in its appeal to those who could solder (which eliminated most good programmers) and to those who could program in machine language (which eliminated most good solderers).

BASIC was generally recognized as the easiest programming language to learn in 1975. It automatically converted simple English-like commands to machine language, effectively removing the programming limitation and at least doubling the number of prospective Altair customers.

Since they didn't have an Altair 8800 computer (nobody did yet), Gates and Allen wrote a program that made a PDP-10 minicomputer at the Harvard Computation Center simulate the Altair's Intel 8080 microprocessor. In six weeks, they wrote a version of the BASIC programming language that would run on the phantom Altair synthesized in the minicomputer. They hoped it would run on a real Altair equipped with at least 4096 bytes of random access memory. The first time they tried to run the language on a real microcomputer was when Paul Allen demonstrated the product to MITS founder Ed Roberts at the company's headquarters in Albuquerque. To their surprise and relief, it worked.

MITS BASIC, as it came to be called, gave substance to the microcomputer. Big computers ran BASIC. Real programs had been written in the language and were performing business, educational, and scientific functions in the real world. While the Altair was a computer of limited power, the fact that Allen and Gates were able to make a high-level language like BASIC run on the platform meant that potential users could imagine running these same sorts of applications now on a desktop rather than on a mainframe.

MITS BASIC was dramatic in its memory efficiency and made the bold move of adding commands that allowed programmers to control the computer memory directly. MITS BASIC wasn't perfect. The authors of the original BASIC, John Kemeny and Thomas Kurtz, both of Dartmouth College, were concerned that Gates and Allen's version deviated from the language they had designed and placed into the public domain a decade before. Kemeny and Kurtz might have been unimpressed, but the hobbyist world was euphoric.

I've got to point out here that for many years Kemeny was president of Dartmouth, a school that didn't accept me when I was applying to colleges. Later, toward the end of the Age of Jimmy Carter, I found myself working for Kemeny, who was then head of the presidential commission investigating the Three Mile Island nuclear accident. One day I told him how Dartmouth had rejected me, and he said, "College admissions are never perfect, though in your case I'm sure we did the right thing." After that I felt a certain affection for Bill Gates.

Gates dropped out of Harvard, Allen left his programming job at Honeywell, and both moved to New Mexico to be close to their customer, in the best Tom Peters style. Hobbyists don't move across country to maintain business relationships, but businessmen do. They camped out in the Sundowner Motel on Route 66 in a neighborhood noted for all-night coffee shops, hookers, and drug dealers.

Gates and Allen did not limit their interest to MITS. They wrote versions of BASIC for other microcomputers as they came to market, leveraging their core technology. The two eventually had a falling out with Ed Roberts of MITS, who claimed that he owned MITS BASIC and its derivatives; they fought and won, something that hackers rarely bothered to do. Capitalists to the bone, they railed against software piracy before it even had a name, writing whining letters to early PC publications.

Gates and Allen started Microsoft with a stated mission of putting "a computer on every desk and in every home, running Microsoft software." Although it seemed ludicrous at the time, they meant it.

While Allen and Gates deliberately went about creating an industry and then controlling it, they were important exceptions to the general trend of PC entrepreneurism. Most of their eventual competitors were people who managed to be in just the right place at the right time and more or less fell into business. These people were mainly enthusiasts who at first developed computer languages and operating systems for their own use. It was worth the effort if only one person—the developer himself—used their product. Often they couldn't even imagine why anyone else would be interested.

Gary Kildall, for example, invented the first microcomputer operating system because he was tired of driving to work. In the early 1970s, Kildall taught computer science at the Naval Postgraduate School in Monterey, California, where his specialty was compiler design. Compilers are software tools that take entire programs written in a high-level language like FORTRAN or Pascal and translate them into assembly language, which can be read directly by the computer. High-level languages are easier to learn than Assembler, so compilers allowed programs to be completed faster and with more features, although the final code was usually longer than if the program had been written directly in the internal language of the microprocessor. Compilers translate, or compile, large sections of code into Assembler at one time, as opposed to interpreters, which translate commands one at a time.

By 1974, Intel had added the 8008 and 8080 to its family of microprocessors and had hired Gary Kildall as a consultant to write software to emulate the 8080 on a DEC time-sharing system, much as Gates and Allen would shortly do at Harvard. Since there were no microcomputers yet, Intel realized that the best way for companies to develop software for microprocessor-based devices was by using such an emulator on a larger system.

Kildall's job was to write the emulator, called Interp/80, followed by a high-level language called PL/M, which was planned as a microcomputer equivalent of the XPL language developed for mainframe computers at Stanford University. Nothing so mundane (and useful by mere mortals) as BASIC for Gary Kildall, who had a Ph.D. in compiler design.

What bothered Kildall was not the difficulty of writing the software but the tedium of driving the fifty miles from his home in Pacific Grove across the Santa Cruz mountains to use the Intel minicomputer in Silicon Valley. He could have used a remote teletype terminal at home, but the terminal was incredibly slow for inputting thousands of lines of data over a phone line; driving was faster.

Or he could develop software directly on the 8080 processor, bypassing the time-sharing system completely. Not only could he avoid the long drive, but developing directly on the microprocessor would also bypass any errors in the minicomputer 8080 emulator. The only problem was that the 8080 microcomputer Gary Kildall wanted to take home didn't exist.

What did exist was the Intellec-8, an Intel product that could be used (sort of) to program an 8080 processor. The Intellec-8 had a microprocessor, some memory, and a port for attaching a Teletype 33 terminal. There was no software and no method for storing data and programs outside of main memory.

The primary difference between the Intellec-8 and a microcomputer was external data storage and the software to control it. IBM had invented a new device, called a floppy disk, to replace punched cards for its minicomputers. The disks themselves could be removed from the drive mechanism, were eight inches in diameter, and held the equivalent of thousands of pages of data. Priced at around $500, the floppy disk drive was perfect for Kildall's external storage device. Kildall, who didn't have $500, convinced Shugart Associates, a floppy disk drive maker, to give him a worn-out floppy drive used in its 10,000-hour torture test. While his friend John Torode invented a controller to link the Intellec-8 and the floppy disk drive, Kildall used the 8080 emulator on the Intel time-sharing system to develop his operating system, called CP/M, or Control Program/Monitor.

If a computer acquires a personality, it does so from its operating system. Users interact with the operating system, which interacts with the computer. The operating system controls the flow of data between a computer and its long-term storage system. It also controls access to system memory and keeps those bits of data that are thrashing around the microprocessor from thrashing into each other. Operating systems usually store data in files, which have individual names and characteristics and can be called up as a program or the user requires them.

Gary Kildall developed CP/M on a DEC PDP-10 minicomputer running the TOPS-10 operating system. Not surprisingly, most CP/M commands and file naming conventions look and operate like their TOPS-10-counterparts. It wasn't pretty, but it did the job.

By the time he'd finished writing the operating system, Intel didn't want CP/M and had even lost interest in Kildall's PL/M language. The only customers for CP/M in 1975 were a maker of intelligent terminals and Lawrence Livermore Labs, which used CP/M to monitor programs on its Octopus network.

In 1976, Kildall was approached by Imsai, the second personal computer company with a stupid name. Imsai manufactured an early 8080-based microcomputer that competed with the Altair. In typical early microcomputer company fashion, Imsai had sold floppy disk drives to many of its customers, promising to send along an operating system eventually. With each of them now holding at least $1,000 worth of hardware that was only gathering dust, the customers wanted their operating system, and CP/M was the only operating system for Intel-based computers that was actually available.

By the time Imsai came along, Kildall and Torode had adapted CP/M to four different floppy disk controllers. There were probably 100 little companies talking about doing 8080-based computers, and neither man wanted to invest the endless hours of tedious coding required to adapt CP/M to each of these new platforms. So they split the parts of CP/M that interfaced with each new controller into a separate computer code module, called the Basic Input/Output System, or BIOS. With all the hardware-dependent parts of CP/M concentrated in the BIOS, it became a relatively easy job to adapt the operating system to many different Intel-based microcomputers by modifying just the BIOS.

With his CP/M and invention of the BIOS, Gary Kildall defined the microcomputer. Peek into any personal computer today, and you'll find a general-purpose operating system adapted to specific hardware through the use of a BIOS, which is now a specialized type of memory chip.

In the six years after Imsai offered the first CP/M computer, more than 500,000 CP/M computers were sold by dozens of makers. Programmers began to write CP/M applications, relying on the operating system's features to control the keyboard, screen, and data storage. This base of applications turned CP/M into a de facto standard among microcomputer operating systems, guaranteeing its long-term success. Kildall started a company called Intergalactic Digital Research (later, just Digital Research) to sell the software in volume to computer makers and direct to users for $70 per copy. He made millions of dollars, essentially without trying.

Before he knew it, Gary Kildall had plenty of money, fast cars, a couple of airplanes, and a business that made increasing demands on his time. His success, while not unwelcome, *was* unexpected, which also meant that it was unplanned for. Success brings with it a whole new set of problems, as Gary Kildall discovered. You can plan for failure, but how do you plan for success?

Every entrepreneur has an objective, which, once achieved, leads to a crisis. In Gary Kildall's case, the objective—just to write CP/M, not even to sell it—was very low, so the crisis came quickly. He was a code god, a programmer who literally saw lines of code fully formed in his mind and then committed them effortlessly to the keyboard in much the same way that Mozart wrote music. He was one with the machine; what did he need with seventy employees?

"Gary didn't give a shit about the business. He was more interested in getting laid," said Gordon Eubanks, a former student of Kildall who led development of computer languages at Digital Research. "So much went so well for so long that he couldn't imagine it would change. When it did—when change was forced upon him—Gary didn't know how to handle it."

"Gary and Dorothy [Kildall's wife and a Digital Research vice-president] had arrogance and cockiness but no passion for products. No one wanted to make the products great. Dan

Bricklin [another PC software pioneer—read on] sent a document saying what should be fixed in CP/M, but it was ignored. Then I urged Gary to do a BASIC language to bundle with CP/M, but when we finally got him to do a language, he insisted on PL/1 —a virtually unmarketable language."

Digital Research was slow in developing a language business to go with its operating systems. It was also slow in updating its core operating system and extending it into the new world of 16-bit microprocessors that came along after 1980. The company in those days was run like a little kingdom, ruled by Gary and Dorothy Kildall.

"In one board meeting," recalled a former Digital Research executive, "we were talking about whether to grant stock options to a woman employee. Dorothy said, 'No, she doesn't deserve options—she's not professional enough; her kids visit her at work after 5:00 P.M.' Two minutes later, Christy Kildall, their daughter, burst into the boardroom and dragged Gary off with her to the stable to ride horses, ending the meeting. Oh yeah, Dorothy knew about professionalism."

Let's say for a minute that Eubanks was correct, and Gary Kildall didn't give a shit about the business. Who said that he had to? CP/M was his invention; Digital Research was his company. The fact that it succeeded beyond anyone's expectations did not make those earlier expectations invalid. Gary Kildall's ambition was limited, something that is not supposed to be a factor in American business. If you hope for a thousand and get a million, you are still expected to want more, but he didn't.

It's easy for authors of business books to get rankled by characters like Gary Kildall who don't take good care of the empires they have built. But in fact, there are no absolute rules of behavior for companies like Digital Research. The business world is, like computers, created entirely by people. God didn't come down and say there will be a corporation and it will have a board of directors. We made that up. Gary Kildall made up Digital Research.

Eubanks, who came to Digital Research after a naval career spent aboard submarines, hated Kildall's apparent lack of discipline, not understanding that it was just a different kind of discipline. Kildall was into programming, not business.

"Programming is very much a religious experience for a lot of people," Kildall explained. "If you talk about programming to a group of programmers who use the same language, they can become almost evangelistic about the language. They form a tight-knit community, hold to certain beliefs, and follow certain rules in their programming. It's like a church with a programming language for a bible."

Gary Kildall's bible said that writing a BASIC compiler to go with CP/M might be a shrewd business move, but it would be a step backward technically. Kildall wanted to break new ground, and a BASIC had already been done by Microsoft.

"The unstated rule around Digital Reseach was that Microsoft did languages, while we did operating systems," Eubanks explained. "It was never stated emphatically, but I always thought that Gary assumed he had an agreement with Bill Gates about this separation and that as long as we didn't compete with Microsoft, they wouldn't compete with us."

Sure.

⌢⌣

The Altair 8800 may have been the first microcomputer, but it was not a commercial success. The problem was that assembly took from forty to an infinite number of hours, depending on the hobbyist's mechanical ability. When the kit was done, the microcomputer either worked or didn't. If it worked, the owner had a programmable computer with a BASIC interpreter, ready to run any software he felt like writing.

The first microcomputer that *was* a major commercial success was the Apple II. It succeeded because it was the first microcomputer that looked like a consumer electronic product. You could buy the Apple from a dealer who would fix it if it broke and would give you at least a little help in learning to operate the beast. The Apple II had a floppy disk drive for data storage, did not require a separate Teletype or video terminal, and offered color graphics in addition to text. Most important, you could buy software written by others that would run on the Apple and with which a novice could do real work.

The Apple II still defines what a low-end computer is like. Twenty-third century archaeologists excavating some ancient ComputerLand stockroom will see no significant functional difference between an Apple II of 1978 and an IBM PS/2 of 1992. Both have processor, memory, storage, and video graphics. Sure, the PS/2 has a faster processor, more memory and storage, and higher-resolution graphics, but that only matters to us today. By the twenty-third century, both machines will seem equally primitive.

The Apple II was guided by three spirits. Steve Wozniak invented the earlier Apple I to show it off to his friends in the Homebrew Computer Club. Steve Jobs was Wozniak's younger sidekick who came up with the idea of building computers for sale and generally nagged Woz and others until the Apple II was working to his satisfaction. Mike Markkula was the semiretired Intel veteran (and one of Noyce's boys) who brought the money and status required for the other two to be taken at all seriously.

Wozniak made the Apple II a simple machine that used clever hardware tricks to get good performance at a smallish price (at least to produce—the retail price of a fully outfitted Apple II was around $3,000). He found a way to allow the microprocessor and the video display to share the same memory. His floppy disk controller, developed during a two-week period in December 1977, used less than a quarter the number of integrated circuits required by other controllers at the time. The Apple's floppy disk controller made it clearly superior to machines appearing about the same time from Commodore and Radio Shack. More so than probably any other microcomputer, the Apple II was the invention of a single person; even Apple's original BASIC interpreter, which was always available in read-only memory, had been written by Woz.

Woz made the Apple II a color machine to prove that he could do it and so he could use the computer to play a color version of Breakout, a video game that he and Jobs had designed for Atari. Markkula, whose main contributions at Intel had been in finance, pushed development of the floppy disk drive so the computer could be used to run accounting programs and store resulting financial data for small business owners. Each man saw the Apple II as a new way of fulfilling an established need— to replace a video game for Woz and a mainframe for Markkula. This followed the trend that new media tend to imitate old media.

Radio began as vaudeville over the air, while early television was radio with pictures. For most users (though not for Woz) the microcomputer was a small mainframe, which explained why Apple's first application for the machine was an accounting package and the first application supplied by a third-party developer was a database—both perfect products for a mainframe substitute. But the Apple II wasn't a very good mainframe replacement. The fact is that new inventions often have to find uses of their own in order to find commercial success, and this was true for the Apple II, which became successful strictly as a spreadsheet machine, a function that none of its inventors visualized.

At $3,000 for a fully configured system, the Apple II did not have a big future as a home machine. Old-timers like to reminisce about the early days of Apple when the company's computers were affordable, but the truth is that they never were.

The Apple II found its eventual home in business, answering the prayers of all those middle managers who had not been able to gain access to the company's mainframe or who were tired of waiting the six weeks it took for the computer department to prepare a report, dragging the answers to simple business questions from corporate data. Instead, they quickly learned to use a spreadsheet program called VisiCalc, which was available at first only on the Apple II.

VisiCalc was a compelling application—an application so important that it, alone justified the computer purchase. Such an application was the last element required to turn the microcomputer from a hobbyist's toy into a business machine. No matter how powerful and brilliantly designed, no computer can be successful without a compelling application. To the people who bought them, mainframes were really inventory machines or accounting machines, and minicomputers were office automation machines. The Apple II was a VisiCalc machine.

VisiCalc was a whole new thing, an application that had not appeared before on some other platform. There were no minicomputer or mainframe spreadsheet programs that could be downsized to run on a microcomputer. The microcomputer and the spreadsheet came along at the same time. They were made for each other.

VisiCalc came about because its inventor, Dan Bricklin, went to business school. And Bricklin went to business school because he thought that his career as a programmer was about to end; it was becoming so easy to write programs that Bricklin was convinced there would eventually be no need for programmers at all, and he would be out of a job. So in the fall of 1977, 26 years old and worried about being washed up, he entered the Harvard Business School looking toward a new career.

At Harvard, Bricklin had an advantage over other students. He could whip up BASIC programs on the Harvard time-sharing system that would perform financial calculations. The problem with Bricklin's programs was that they had to be written and rewritten for each new problem. He began to look for a more general way of doing these calculations in a format that would be flexible.

What Bricklin really wanted was not a microcomputer program at all but a specialized piece of hardware—a kind of very advanced calculator with a heads-up display similar to the weapons system controls on an F-14 fighter. Like Luke Skywalker jumping into the turret of the *Millennium Falcon*, Bricklin saw himself blasting out financials, locking onto profit and loss numbers that would appear suspended in space before him. It was to be a business tool cum video game, a Saturday Night Special for M.B.A.s, only the hardware technology didn't exist in those days to make it happen.

Back in the semireal world of the Harvard Business School, Bricklin's production professor described large blackboards that were used in some companies for production planning. These blackboards, often so long that they spanned several rooms, were segmented in a matrix of rows and columns. The production planners would fill each space with chalk scribbles relating to the time, materials, manpower, and money needed to manufacture a product. Each cell on the blackboard was located in both a column and a row, so each had a two-dimensional address. Some cells were related to others, so if the number of workers listed in cell C-3 was increased, it meant that the amount of total wages in cell D-5 had to be increased proportionally, as did the total number of items produced, listed in cell F-7. Changing the value in one cell required the recalculation of values in all other linked cells, which took a lot of erasing and a lot of recalculating and left the planners constantly worried that they had overlooked recalculating a linked value, making their overall conclusions incorrect.

Given that Bricklin's Luke Skywalker approach was out of the question, the blackboard metaphor made a good structure for Bricklin's financial calculator, with a video screen replacing the physical blackboard. Once data and formulas were introduced by the user into each cell, changing one variable would automatically cause all the other cells to be recalculated and changed too. No linked cells could be forgotten. The video screen would show a window on a spreadsheet that was actually held in computer memory. The virtual spreadsheet inside the box could be almost any size, putting on a desk what had once taken whole rooms filled with blackboards. Once the spreadsheet was set up, answering a what-if question like "How much more money will we make if we raise the price of each widget by a dime?" would take only seconds.

His production professor loved the idea, as did Bricklin's accounting professor. Bricklin's finance professor, who had others to do his computing for him, said there were already financial analysis programs running on mainframes, so the world did not need Dan Bricklin's little program. Only the world *did* need Dan Bricklin's little program, which still didn't have a name.

It's not surprising that VisiCalc grew out of a business school experience because it was the business schools that were producing most of the future VisiCalc users. They were the thousands of M.B.A.s who were coming into the workplace trained in analytical business techniques and, even more important, in typing. They had the skills and the motivation but usually not the access to their company computer. They were the first generation of businesspeople who could do it all by themselves, given the proper tools.

Bricklin cobbled up a demonstration version of his idea over a weekend. It was written in BASIC, was slow, and had only enough rows and columns to fill a single screen, but it demonstrated many of the basic functions of the spreadsheet. For one thing, it just sat there. This is the genius of the spreadsheet; it's event driven. Unless the user changes a cell, nothing happens. This may not seem like much, but being event driven makes a spreadsheet totally responsive to the user; it puts the user in charge in a way that most other programs did not. VisiCalc was a spreadsheet language, and what the users were doing was rudimentary programming, without the anxiety of knowing that's what it was.

By the time Bricklin had his demonstration program running, it was early 1978 and the mass market for microcomputers, such as it was, was being vied for by the Apple II, Commodore PET, and the Radio Shack TRS-80. Since he had no experience with micros, and so no preference for any particular machine, Bricklin and Bob Frankston, his old friend from MIT and new partner, developed VisiCalc for the Apple II, strictly because that was the computer their would-be publisher loaned them in the fall of 1978. No technical merit was involved in the decision.

Dan Fylstra was the publisher. He had graduated from Harvard Business School a year or two before and was trying to make a living selling microcomputer chess programs from his home. Fylstra's Personal Software was the archetypal microcomputer application software company. Bill Gates at Microsoft and Gary Kildall at Digital Research were specializing in operating systems and languages, products that were lumped together under the label of systems software, and were mainly sold to hardware manufacturers rather than directly to users. But Fylstra was selling applications direct to retailers and end users, often one program at a time. With no clear example to follow, he had to make most of the mistakes himself, and did.

Since there was no obvious success story to emulate, no retail software company that had already stumbled across the rules for making money, Fylstra dusted off his Harvard case study technique and looked for similar industries whose rules could be adapted to the microcomputer software biz. About the closest example he could find was book publishing, where the author accepts responsibility for designing and implementing the product, and the publisher is responsible for manufacturing, distribution, marketing, and sales. Transferred to the microcomputer arena, this meant that Software Arts, the company Bricklin and Frankston formed, would develop VisiCalc and its subsequent versions, while Personal Software, Fylstra's company, would copy the floppy disks, print the manuals, place ads in computer publications, and distribute the product to retailers and the public. Software Arts would receive a royalty of 37.5 percent on copies of VisiCalc sold at retail and 50 percent for copies sold wholesale. "The numbers seemed fair at the time," Fylstra said.

Bricklin was still in school, so he and Frankston divided their efforts in a way that would become a standard for microcomputer programming projects. Bricklin designed the program, while Frankston wrote the actual code. Bricklin would say, "This is the way the program is supposed to look, these are the features, and this is the way it should function," but the actual design of the internal program was left up to Bob Frankston, who had been writing software since 1963 and was clearly up to the task. Frankston added a few features on his own, including one called "lookup," which could extract values from a table, so he could use VisiCalc to do his taxes.

Bob Frankston is a gentle man and a brilliant programmer who lives in a world that is just slightly out of sync with the world in which you and I live. (Okay, so it's out of sync with the world in which *you* live.) When I met him, Frankston was chief scientist at Lotus Development, the people who gave us the Lotus 1-2-3 spreadsheet. In a personal computer hardware or software company, being named chief scientist means that the boss doesn't know what to do with you. Chief scientists don't generally have to *do* anything; they're just smart people whom the company doesn't want to lose to a competitor. So they get a title and an office and are obliged to represent the glorious past at all company functions. At Apple Computer, they call them Apple Fellows, because you can't have more than one chief scientist.

Bob Frankston, a modified nerd (he combined the requisite flannel shirt with a full beard), seemed not to notice that his role of chief scientist was a sham, because to him it wasn't; it was the perfect opportunity to look inward and think deep thoughts without regard to their marketability.

"Why are you doing this as a book?" Frankston asked me over breakfast one morning in Newton, Massachusetts. By "this," he meant the book you have in your hands right now, the major literary work of my career and, I hope, the basis of an important American fortune. "Why not do it as a hypertext file that people could just browse through on their computers?"

I will not be browsed through. The essence of writing books is the author's right to tell the story in his own words and in the order he chooses. Hypertext, which allows an instant accounting of how many times the words *Dynamic Random-Access Memory* or *fuck* appear, completely eliminates what I perceive as my value-added, turns this exercise into something like the Yellow Pages, and totally eliminates the prospect that it will help fund my retirement.

"Oh," said Frankston, with eyebrows raised. "Okay."

Meanwhile, back in 1979, Bricklin and Frankston developed the first version of VisiCalc on an Apple II emulator running on a minicomputer, just as Microsoft BASIC and CP/M had been written. Money was tight, so Frankston worked at night, when computer time was cheaper and when the time-sharing system responded faster because there were fewer users.

They thought that the whole job would take a month, but it took close to a year to finish. During this time, Fylstra was showing prerelease versions of the product to the first few software retailers and to computer companies like Apple and Atari. Atari was interested but did not yet have a computer to sell. Apple's reaction to the product was lukewarm.

VisiCalc hit the market in October 1979, selling for $100. The first 100 copies went to Marv Goldschmitt's computer store in Bedford, Massachusetts, where Dan Bricklin appeared regularly to give demonstrations to bewildered customers. Sales were slow. Nothing like this product had existed before, so it would be a mistake to blame the early microcomputer users for not realizing they were seeing the future when they stared at their first VisiCalc screen.

Nearly every software developer in those days believed that small businesspeople would be the main users of any financial products they'd develop. Markkula's beloved accounting system, for example, would be used by small retailers and manufacturers who could not afford access to a time-sharing system and preferred not to farm the job out to an accounting service. Bricklin's spreadsheet would be used by these same small businesspeople to prepare budgets and forecast business trends. Automation was supposed to come to the small business community through the microcomputer just as it had come to the large and medium businesses through mainframes and minicomputers. But it didn't work that way.

The problem with the small business market was that small businesses weren't, for the most part, very businesslike. Most small businesspeople didn't know what they were doing. Accounting was clearly beyond them.

At the time, sales to hobbyists and would-be computer game

players were topping out, and small businesses weren't buying. Apple and most of its competitors were in real trouble. The personal computer revolution looked as if it might last only five years. But then VisiCalc sales began to kick in.

Among the many customers who watched VisiCalc demos at Marv Goldschmitt's computer store were a few businesspeople—rare members of both the set of computer enthusiasts and the economic establishment. Many of these people had bought Apple IIs, hoping to do real work until they attempted to come to terms with the computer's forty-column display and lack of lowercase letters. In VisiCalc, they found an application that did not care about lowercase letters, and since the program used a view through the screen on a larger, virtual spreadsheet, the forty-column limit was less of one. For $100, they took a chance, carried the program home, then eventually took both the program and the computer it ran on with them to work. The true market for the Apple II turned out to be big business, and it was through the efforts of enthusiast employees, not Apple marketers, that the Apple II invaded industry.

"The beautiful thing about the spreadsheet was that customers in big business were really smart and understood the benefits right away," said Trip Hawkins, who was in charge of small business strategy at Apple. "I visited Westinghouse in Pittsburgh. The company had decided that Apple II technology wasn't suitable, but 1,000 Apple IIs had somehow arrived in the corporate headquarters, bought with petty cash funds and popularized by the office intelligentsia."

Hawkins was among the first to realize that the spreadsheet was a new form of computer life and that VisiCalc—the only spreadsheet on the market and available at first only on the Apple II—would be Apple's tool for entering, maybe dominating, the microcomputer market for medium and large corporations. VisiCalc was a strategic asset and one that had to be tied up fast before Bricklin and Frankston moved it onto other platforms like the Radio Shack TRS-80.

"When I brought the first copies of VisiCalc into Apple, it was clear to me that this was an important application, vital to the success of the Apple II," Hawkins said. "We didn't want it to appear on the Radio Shack or on the IBM machine we knew was coming, so I took Dan Fylstra to lunch and talked about a buy-out. The price we settled on would have been $1 million worth of Apple stock, which would have been worth much more later. But when I took the deal to Markkula for approval, he said, 'No, it's too expensive.'"

A million dollars was an important value point in the early microcomputer software business. Every programmer who bothered to think about money at all looked toward the time when he would sell out for a cool million. Apple could have used owner-ship of the program to dominate business microcomputing for years. The deal would have been good, too, for Dan Fylstra, who so recently had been selling chess programs out of his apartment. Except that Dan Fylstra didn't own VisiCalc—Dan Bricklin and Bob Frankston did. The deal came and went without the boys in Massachusetts even being told.

ROLE MODELS

This being the 1990s, the economy is shot to hell and we've got nothing much better to do, the personal computer industry is caught up in an issue called *look and feel*, which means that your computer software can't look too much like my computer software or I'll take you to court. Look and feel is a matter of not only how many angels can dance on the head of a pin but what dance it is they are doing and who owns the copyright.

Here's an example of look and feel. It's 1913, and we're at the Notre Dame versus Army football game (this is all taken straight from the film *Knute Rockne, All-American*, in which young Ronald Reagan appeared as the ill-fated Gipper—George Gipp). Changing football forever, the Notre Dame quarterback throws the first-ever forward pass, winning the game. A week later, Notre Dame is facing another team, say Purdue. By this time, word of the forward pass has gotten around, the Boilermakers have thrown a few in practice, and they like the effect. So early in the first quarter, the Purdue quarterback throws a forward pass. The Notre Dame coach calls a time-out and sends young Knute Rockne jogging over to the Purdue bench.

"Coach says that'll be five dollars," mumbles an embarrassed Knute, kicking at the dirt with his toe.

"Say what, son?"

"Coach says the forward pass is Notre Dame property, and if you're going to throw one, you'll have to pay us five dollars. I can take a check."

That's how it works. Be the first one on your block to invent a particular way of doing something, and you can charge the world for copying your idea or even prohibit the world from copying it at all. It doesn't even matter if the underlying mechanism is different; if the two techniques look similar, one is probably violating the look and feel of the other.

Just think of the money that could have been earned by the first person to put four legs on a chair or to line up the clutch, brake, and gas pedals of a car in that particular order. My secret suspicion is that this sort of easy money was the real reason Alexander Graham Bell tried to get people to say "ahoy" when they answered the telephone rather than "hello." It wasn't enough that he was getting a nickle for the phone call; Bell wanted another nickle for his user interface.

There's that term, *user interface*. User interface is at the heart of the look-and-feel debate because it's the user interface that we're always looking at and feeling. Say the Navajo nation wants to get back to its computing roots by developing a computer system that uses smoke signals to transfer data into and out of the computer. Whatever system of smoke puffs they settle on will be that computer's user interface, and therefore protectable under law (U.S., not tribal).

What's particularly ludicrous about this look-and-feel business is that it relies on all of us believing that there is something uniquely valuable about, for example, Apple Computer's use of overlapping on-screen windows or pull-down menus. We are supposed to pretend that some particular interface concepts sprang fully grown and fully clothed from the head of a specific programmer, totally without reference to prior art.

Bullshit.

Nearly everything in computing, both inside and outside the box, is derived from earlier work. In the days of mainframes and minicomputers and early personal computers like the Apple II and the Tandy TRS-80, user interfaces were based on the mainframe model of typing out commands to the computer one 80-character line at a time—the same line length used by punch cards (IBM should have gone for the bucks with that one but didn't). But the commands were so simple and obvious that it seemed stupid at the time to view them as proprietary. Gary Kildall stole his command set for CP/M from DEC's TOPS-10 minicomputer operating system, and DEC never thought to ask for a dime. Even IBM's VM command set for mainframe computers was copied by a PC operating system called Oasis (now Theos), but IBM probably never even noticed.

This was during the first era of microcomputing, which lasted from the introduction of the MITS Altair 8800 in early 1975 to the arrival of the IBM Personal Computer toward the end of 1981. Just like the golden age of television, it was a time when technology was primitive and restrictions on personal creativity were minimal, too, so everyone stole from everyone else. This was the age of 8-bit computing, when Apple, Commodore, Radio Shack, and a hundred-odd CP/M vendors dominated a small but growing market with their computers that processed data eight bits at a time. The flow of data through these little computers was like an eight-lane highway, while the minicomputers and mainframes had their traffic flowing on thirty-two lanes and more. But eight lanes were plenty, considering what the Apples and others were trying to accomplish, which was to put the computing environment of a mainframe computer on a desk for around $3,000.

Mainframes weren't that impressive. There were no fancy, high-resolution color graphics in the mainframe world—nothing that looked even as good as a television set. Right from the beginning, it was possible to draw pictures on an Apple II that were impossible to do on an IBM mainframe.

Today, for example, several million people use their personal computers to communicate over worldwide data networks, just for fun. I remember when a woman on the CompuServe network ran a nude photo of herself through an electronic scanner and sent the digitized image across the network to all the men with whom she'd been flirting for months on-line. In grand and glorious high-resolution color, what was purported to be her yummy flesh scrolled across the screens of dozens of salivating computer nerds, who quickly forwarded the image to hundreds and then thousands of their closest friends. You couldn't send such an image from one terminal to another on a mainframe computer; the technology doesn't exist, or all those wacky secretaries who have hopped on Xerox machines to photocopy their backsides would have had those backsides in electronic distribution years ago.

My point is that the early pioneers of microcomputing stole freely from the mainframe and minicomputer worlds, but there wasn't really much worth stealing, so nobody was bothered. But with the introduction of 16-bit microprocessors in 1981 and 1982, the mainframe role model was scrapped altogether. This second era of microcomputing required a new role model and new ideas to copy. And this time around, the ideas were much more powerful—so powerful that they were worth protecting, which has led us to this look-and-feel fiasco. Most of these new ideas came from the Xerox Palo Alto Research Center (PARC). They still do.

To understand the personal computer industry, we have to understand Xerox PARC, because that's where the most of the computer technology that we'll use for the rest of the century was invented.

There are two kinds of research: research and development and basic research. The purpose of research and development is to invent a product for sale. Edison invented the first commercially successful light bulb, but he did not invent the underlying science that made light bulbs possible. Edison at least understood the science, though, which was the primary difference between inventing the light bulb and inventing fire.

The research part of R&D develops new technologies to be used in a specific product, based on existing scientific knowledge. The development part of R&D designs and builds a product using those technologies. It's possible to do development without research, but that requires licensing, borrowing, or stealing research from somewhere else. If research and development is successful, it results in a product that hits the market fairly soon— usually within eighteen to twenty-four months in the personal computer business.

Basic research is something else—ostensibly the search for knowledge for its own sake. Basic research provides the scientific knowledge upon which R&D is later based. Sending telescopes into orbit or building superconducting supercolliders is basic research. There is no way, for example, that the $1.5 billion Hubble space telescope is going to lead directly to a new car or computer or method of solid waste disposal. That's not what it's for.

If a product ever results from basic research, it usually does so fifteen to twenty years down the road, following a later period of research and development.

What basic research is really for depends on who is doing the research and how they are funded. Basic research takes place in government, academic, and industrial laboratories, each for a different purpose. Basic research in government labs is used primarily to come up with new ideas for blowing up the world before someone else in some unfriendly country comes up with those same ideas. While the space telescope and the supercollider are civilian projects intended to explain the nature and structure of the universe, understanding that nature and structure are very important to anyone planning the next generation of earth-shaking weapons. Two thirds of U.S. government basic research is typically conducted for the military, with health research taking most of the remaining funds.

Basic research at universities comes in two varieties: research that requires big bucks and research that requires small bucks. Big bucks research is much like government research and in fact usually *is* government research but done for the government under contract. Like other government research, big bucks academic research is done to understand the nature and structure of the universe or to understand life, which really means that it is either for blowing up the world or extending life, whichever comes first. Again, that's the government's motivation. The universities' motivation for conducting big bucks research is to bring in money to support professors and graduate students and to wax the floors of ivy-covered buildings. While we think they are busy teaching and learning, these folks are mainly doing big bucks basic research for a living, all the while priding themselves on their terrific summer vacations and lack of a dress code.

Small bucks basic research is the sort that requires paper and pencil, and maybe a blackboard, and is aimed primarily at increasing knowledge in areas of study that don't usually attract big bucks—that is, areas that don't extend life or end it, or both. History, political science, and romance languages are typical small bucks areas of basic research. The real purpose of small bucks research to the universities is to provide a means of deciding, by the quality of their small bucks research, which professors in these areas should get tenure.

Nearly all companies do research and development, but only a few do basic research. The companies that can afford to do basic research (and can't afford not to) are ones that dominate their markets. Most basic research in industry is done by companies that have at least a 50 percent market share. They have both the greatest resources to spare for this type of activity and the most to lose if, by choosing not to do basic research, they eventually lose their technical advantage over competitors. Such companies typically devote about 1 percent of sales each year to research intended not to develop specific products but to ensure that the company remains a dominant player in its industry twenty years from now. It's cheap insurance, since failing to do basic research guarantees that the next major advance will be owned by someone else.

The problem with industrial basic research, and what differentiates it from government basic research, is this fact that its true product is insurance, not knowledge. If a researcher at the government-sponsored Lawrence Livermore Lab comes up with some particularly clever new way to kill millions of people, there is no doubt that his work will be exploited and that weapons using the technology will eventually be built. The simple rule about weapons is that if they can be built, they will be built. But basic researchers in industry find their work is at the mercy of the marketplace and their captains-of-industry bosses. If a researcher at General Motors comes up with a technology that will allow cars to be built for $100 each, GM executives will quickly move to bury the technology, no matter how good it is, because it threatens their current business, which is based on cars that cost thousands of dollars each to build. Consumers would revolt if it became known that GM was still charging high prices for cars that cost $100 each to build, so the better part of business valor is to stick with the old technology since it results in more profit dollars per car produced.

In the business world, just because something *can* be built does not at all guarantee that it *will* be built, which explains why RCA took a look at the work of George Heilmeier, a young researcher working at the company's research center in New Jersey and quickly decided to stop work on Heilmeier's invention, the liquid crystal display. RCA made this mid-1960s decision because LCDs might have threatened its then-profitable business of building cathode ray picture tubes. Twenty-five years later, of course, RCA is no longer a factor in the television market, and LCD displays—nearly all made in Japan—are everywhere.

Most of the basic research in computer science has been done at universities under government contract, at AT&T Bell Labs in New Jersey and in Illinois, at IBM labs in the United States, Europe, and Japan, and at the Xerox PARC in California. It's PARC that we are interested in because of its bearing on the culture of the personal computer.

Xerox PARC was started in 1970 when leaders of the world's dominant maker of copying machines had a sinking feeling that paper was on its way out. If people started reading computer screens instead of paper, Xerox was in trouble, *unless* the company could devise a plan that would lead it to a dominant position in the paperless office envisioned for 1990. That plan was supposed to come from Xerox PARC, a group of very smart people working in buildings on Coyote Hill Road in the Stanford Industrial Park near Stanford University.

The Xerox researchers were drawn together over the course of a few months from other corporations and from universities and then plunked down in the golden hills of California, far from any other Xerox facility. They had nothing at all to do with copiers, yet they worked for a copier company. If they came to have a feeling of solidarity, then, it was much more with each other than with the rest of Xerox. The researchers at PARC soon came to look down on the marketers of Xerox HQ, especially when they were asked questions like, "Why don't you do all your programming in BASIC—it's so much easier to learn," which was like suggesting that Yehudi Menuhin switch to rhythm sticks.

The researchers at PARC were iconoclastic, independent, and not even particularly secretive, since most of their ideas would not turn into products for decades. They became the celebrities of computer science and were even profiled in *Rolling Stone*.

PARC was supposed to plot Xerox a course into the electronic office of the 1990s, and the heart of that office would be, as it always had been, the office worker. Like designers of typewriters and adding machines, the deep thinkers at Xerox PARC had to develop systems that would be useful to lightly trained people working in an office. This is what made Xerox different from every other computer company at that time.

Some of what developed as the PARC view of future computing was based on earlier work by Doug Engelbart, who worked at the Stanford Research Institute in nearby Menlo Park. Engelbart was the first computer scientist to pay close attention to user interface—how users interact with a computer system. If computers could be made easier to use, Engelbart thought, they would be used by more people and with better results.

Punch cards entered data into computers one card at a time. Each card carried a line of data up to 80 characters wide. The first terminals simply replaced the punch card reader with a new input device; users still submitted data, one 80-column line at a time, through a computer terminal. While the terminal screen might display as many as 25 lines at a time, only the bottom line was truly active and available for changes. Once the carriage return key was punched, those data were *in* the computer: no going back to change them later, at least not without telling the computer that you wanted to reedit line 32, please.

I once wrote an entire book using a line editor on an IBM mainframe, and I can tell you it was a pain.

Engelbart figured that real people in real offices didn't write letters or complete forms one line at a time, with no going back. They thought in terms of pages, rather than lines, and their pens and typewriters could be made to scroll back and forth and move vertically on the page, allowing access to any point. Engelbart wanted to bring that page metaphor to the computer by inventing a terminal that would allow users to edit anywhere on the screen. This type of terminal required some local intelligence, keeping the entire screen image in the terminal's memory. This intelligence was also necessary to manage a screen that was much more flexible than its line-by-line predecessor; it was comprised of thousands of points that could be turned on or off.

The new point-by-point screen technology, called bit mapping, also required a means for roaming around the screen. Engelbart used what he called a mouse, which was a device the size of a pack of cigarettes on wheels that could be rolled around on the table next to the terminal and was connected to the terminal by a wire. Moving the mouse caused the cursor on-screen to move too.

With Engelbart's work as a start, the folks at PARC moved toward prototyping more advanced systems of networked computers that used mice, page editors, and bit-mapped screens to make computing easier and more powerful.

During the 1970s, the Computer Science Laboratory (CSL) at Xerox PARC was the best place in the world for doing computer research. Researchers at PARC invented the first high-speed computer networks and the first laser printers, and they devised the first computers that could be called easy to use, with intuitive graphical displays. The Xerox Alto, which had built-in networking, a black-on-white bit-mapped screen, a mouse, and hard disk data storage and sat under the desk looking like R2D2, was the most sophisticated computer workstation of its time, because it was the *only* workstation of its time. Like the other PARC advances, the Alto was a wonder, but it wasn't a product. Products would have taken longer to develop, with all their attendant questions about reliability, manufacturability, marketability, and profitability—questions that never once crossed a brilliant mind at PARC. Nobody was expected to buy computers built by Xerox PARC.

There is a very good book about Xerox PARC called *Fumbling the Future*, which says that PARC researchers Butler Lampson and Chuck Thacker were inventing the first personal computer when they designed and built the Alto in 1972 and 1973 and that by choosing not to commercialize the Alto, Xerox gave up its chance to become the dominant player in the coming personal computer revolution. The book is good, but this conclusion is wrong. Just the parts to build an Alto in 1973 cost $10,000, which suggests that a retail Alto would have had to sell for at least $25,000 (1973 dollars, too) for Xerox to make money on it. When personal computers finally did come along a couple of years later, the price point that worked was around $3,000, so the Alto was way too expensive. It wasn't a personal computer.

And there was no compelling application on the Alto—no VisiCalc, no single function—that could drive a potential user out of the office, down the street, and into a Xerox showroom just to buy it. The idea of a spreadsheet never came to Xerox. Peter Deutsch wrote about what he called spiders—values (like 1989 revenues) that appeared in multiple documents, all linked together. Change a value in one place and the spider made sure that value was changed in all linked places. Spiders were like spreadsheets without the grid of columns and rows and without the clearly understood idea that the linked values were used to solve quantitative problems. Spiders weren't VisiCalc.

If Xerox made a mistake in its handling of the Alto, it was in *almost* choosing to sell it. The techies at PARC knew that the Alto was the best workstation around, but they didn't think about the pricing and application issues. When Xerox toyed with the idea of selling the Alto, that consideration instantly erased any doubts in the minds of its developers that theirs was a commercial system. Dave Kerns, the president of Xerox, kept coming around, nodding his head, and being supportive but somehow never wrote the all-important check.

Xerox's on-again, off-again handling of the Alto alienated the technical staff at PARC, who never really understood why their system was not marketed. To them, it seemed as if Kerns and Xerox, like the owners of Sutter's Mill, had found gold in the stream but decided to build condos on the spot instead of mining because it was never meant to be a gold mine.

There was a true sense of the academic—the amateur—in Ethernet too. PARC's technology for networking all its computers together was developed in 1973 by a team led by Bob Metcalfe. Metcalfe's group was looking for a way to speed up the link between computers and laser printers, both of which had become so fast that the major factor slowing down printing was, in fact, the wire between the two machines rather than anything having to do with either the computer or the printer. The image of the page was created in the memory of the computer and then had to be transmitted bit by bit to the printer. At 600 dots-per-inch resolution, this meant sending more than 33 million bits across the wire for each page. The computer could resolve the page in memory in 1 second and the printer could print the page in 2 seconds, but sending the data over what was then considered to be a high-speed serial link took just under 15 *minutes*. If laser printers were going to be successful in the office, a faster connection would have to be invented.

PARC's printers were computers in their own right that talked back and forth with the computers they were attached to, and this two-way conversation meant that data could collide if both systems tried to talk at once. Place a dozen or more computers and printers on the same wire, and the risk of collisions was even greater. In the absence of a truly great solution to the collision problem, Metcalfe came up with one that was at least truly good and time honored: he copied the telephone party line. Good neighbors listen on their party line first, before placing a call, and that's what Ethernet devices do too—listen, and if another transmission is heard, they wait a random time interval before trying again. Able to transmit data at 2.67 million bits per second across a coaxial cable, Ethernet was a technical triumph, cutting the time to transmit that 600 dpi page from 15 minutes down to 12 seconds.

At 2.67 megabits per second (mbps), Ethernet was a hell of a product, for both connecting computers to printers and, as it turned out, connecting computers to other computers. Every Alto came with Ethernet capability, which meant that each computer had an individual address or name on the network. Each user named his own Alto. John Ellenby, who was in charge of building the Altos, named his machine Gzunda "because it gzunda the desk."

The 2.67 mbps Ethernet technology was robust and relatively simple. But since PARC wasn't supposed to be interested in doing products at all but was devoted instead to expanding the technical envelope, the decision was made to scale Ethernet up to 10 mbps over the same wire with the idea that this would allow networked computers to split tasks and compute in parallel.

Metcalfe had done some calculations that suggested the marketplace would need only 1 mbps through 1990 and 10 mbps through the year 2000, so it was decided to aim straight for the millennium and ignore the fact that 2.67 mbps Ethernet would, by these calculations, have a useful product life span of approximately twenty years. Unfortunately, 10 mbps Ethernet was a much more complex technology—so much more complex that it literally turned what might have been a product back into a technology exercise. Saved from its brush with commercialism, it would be another six years before 10 mbps Ethernet became a viable product, and even then it wouldn't be under the Xerox label.

Beyond the Alto, the laser printer, and Ethernet, what Xerox PARC contributed to the personal computer industry was a way of working—Bob Taylor's way of working.

Taylor was a psychologist from Texas who in the early 1960s got interested in what people could and ought to do with computers. He wasn't a computer scientist but a visionary who came to see his role as one of guiding the real computer scientists in their work. Taylor began this task at NASA and then shifted a couple years later to working at the Department of Defense's Advanced Research Projects Administration (ARPA). ARPA was a brainchild program of the Kennedy years, intended to plunk money into selected research areas without the formality associated with most other federal funding. The ARPA funders, including Taylor, were supposed to have some idea in what direction technology ought to be pushed to stay ahead of the Soviet Union, and they were expected to do that pushing with ARPA research dollars. By 1965, 33-year-old Bob Taylor was in control of the world's largest governmental budget for advanced computer research.

At ARPA, Taylor funded fifteen to twenty projects at a time at companies and universities throughout the United States. He brought the principal researchers of these projects together in regular conferences where they could share information. He funded development of the ARPAnet, the first nationwide com-

puter communications network, primarily so these same researchers could stay in constant touch with each other. Taylor
made it his job to do whatever it took to find the best people
doing the best work and help them to do more.

When Xerox came calling in 1970, Taylor was already out of
the government following an ugly experience reworking U.S.
military computer systems in Saigon during the Vietnam War.
For the first time, Taylor had been sent to solve a real-world computing problem, and reality didn't sit well with him. Better to get
back to the world of ideas, where all that was corrupted were the
data, and there was no such thing as a body count.

Taylor held a position at the University of Utah when Xerox
asked him to work as a consultant, using his contacts to help staff
what was about to become the Computer Science Laboratory
(CSL) at PARC. Since he wasn't a researcher, himself, Taylor
wasn't considered qualified to run the lab, though he eventually
weaseled into that job too.

Alan Kay, jazz musician, computer visionary, and Taylor's
first hire at PARC, liked to say that of the top one hundred computer researchers in the world, fifty-eight of them worked at
PARC. And sometimes he said that seventy-six of the top one hundred worked at PARC. The truth was that Taylor's lab never had
more than fifty researchers, so both numbers were inflated, but it
was also true that for a time under Taylor, CSL certainly worked as
though there were many more than fifty researchers. In less than
three years from its founding in 1970, CSL researchers built their
own time-sharing computer, built the Alto, and invented both the
laser printer and Ethernet.

To accomplish so much so fast, Taylor created a flat organizational structure; everyone who worked at CSL, from scientists to
secretaries, reported directly to Bob Taylor. There were no middle
managers. Taylor knew his limits, though, and those limits said
that he had the personal capacity to manage forty to fifty

researchers and twenty to thirty support staff. Changing the world with that few people required that they all be the best at what they did, so Taylor became an elitist, hiring only the best people he could find and subjecting potential new hires to rigorous examination by their peers, designed to "test the quality of their nervous systems." Every new hire was interviewed by everyone else at CSL. Would-be researchers had to appear in a forum where they were asked to explain and defend their previous work. There were no junior research people. Nobody was wooed to work at CSL; they were challenged. The meek did not survive.

Newly hired researchers typically worked on a couple of projects with different groups within CSL. Nobody worked alone. Taylor was always cross-fertilizing, shifting people from group to group to get the best mix and make the most progress. Like his earlier ARPA conferences, Taylor chaired meetings within CSL where researchers would present and defend their work. These sessions came to be called Dealer Meetings, because they took place in a special room lined with blackboards, where the presenter stood like a blackjack dealer in the center of a ring of beanbag chairs, each occupied by a CSL genius taking potshots at this week's topic. And there was Bob Taylor, too, looking like a high school science teacher and keeping overall control of the process, though without seeming to do so.

Let's not underestimate Bob Taylor's accomplishment in just getting these people to communicate on a regular basis. Computer people love to talk about their work—but *only* their work. A Dealer Meeting not under the influence of Bob Taylor would be something like this:

Nerd A (the dealer): "I'm working on this pattern recognition problem, which I see as an important precursor to teaching computers how to read printed text."

Nerd B (in the beanbag chair): "That's okay, I guess, but I'm working on algorithms for compressing data. Just last night I figured out how to . . . ''

See? Without Taylor it would have been chaos. In the Dealer Meetings, as in the overall intellectual work of CSL, Bob Taylor's function was as a central switching station, monitoring the flow of ideas and work and keeping both going as smoothly as possible. And although he wasn't a computer scientist and couldn't actually do the work himself, Taylor's intermediary role made him so indispensable that it was always clear who worked for whom. Taylor was the boss. They called it "Taylor's lab."

While Bob Taylor set the general direction of research at CSL, the ideas all came from his technical staff. Coming up with ideas and then turning them into technologies was all these people had to do. They had no other responsibilities. While they were following their computer dreams, Taylor took care of everything else: handling budgets, dealing with Xerox headquarters, and generally keeping the whole enterprise on track. And his charges didn't always make Taylor's job easy.

Right from the start, for example, they needed a DEC PDP-10 time-sharing system, because that was what Engelbart had at SRI, and PDP-10s were also required to run the ARPAnet software. But Xerox had its own struggling minicomputer operation, Scientific Data Systems, which was run by Max Palevsky down in El Segundo. Rather than buy a DEC computer, why not buy one of Max's Sigma computers, which competed directly with the PDP-10? Because software is vastly more complex than hardware, that's why. You could build your own copy of a PDP-10 in less time than it would take to modify the software to run on Xerox's machine! And so they did. CSL's first job on their way toward the office of the future was to clone the PDP-10. They built the Multi-Access Xerox Computer (MAXC). The C was silent, just to make sure that Max Palevsky knew the computer was named in his honor.

The way to create knowledge is to start with a strong vision and then ruthlessly abandon parts of that vision to uncover some greater truth. Time sharing was part of the original vision at CSL because it had been part of Engelbart's vision, but having gone to the trouble of building its own time-sharing system, the researchers at PARC soon realized that time sharing itself was part of the problem. MAXC was thrown aside for networks of smaller computers that communicated with each other—the Alto.

Taylor perfected the ideal environment for basic computer research, a setting so near to perfect that it enabled four dozen people to invent much of the computer technology we have today, led not by another computer scientist but by an exceptional administrator with vision.

I'm writing this in 1991, when Bill Gates of Microsoft is traveling the world preaching a new religion he calls Information At Your Fingertips. The idea is that PC users will be able to ask their machines for information, and, if it isn't available locally, the PC will figure out how and where to find it. No need for Joe User to know where or how the information makes its way to his screen. That stuff can be left up to the PC and to the many other systems with which it talks over a network. Gates is making a big deal of this technology, which he presents pretty much as his idea. But *Information At Your Fingertips was invented at Xerox PARC in 1973!* Like so many PARC inventions, though, it's only now that we have the technology to implement it at a price normal mortals can afford.

In its total dedication to the pursuit of knowledge, CSL was like a university, except that the pay and research budgets were higher than those usually found in universities and there was no teaching requirement. There was total dedication to doing the best work with the best people—a purism that bordered on arrogance, though Taylor preferred to see it more as a relentless search for excellence.

What sounded to the rest of the world like PARC arrogance was really the fallout of the lab's intense and introverted intellectual environment. Taylor's geniuses, used to dealing with each other and not particularly sensitive to the needs of mere mortals, thought that the quality of their ideas was self-evident. They didn't see the need to explain—to translate the idea into the world of the other person. Beyond pissing off Miss Manners, the fatal flaw in this PARC attitude was their failure to understand that there were other attributes to be considered as well when examining every idea. While idea A may be, in fact, better than idea B, A is not always cheaper, or more timely, or even possible —factors that had little relevance in the think tank but terrific relevance in the marketplace.

In time the dream at CSL and Xerox PARC began to fade, not because Taylor's geniuses had not done good work but because Xerox chose not to do much with the work they had done. Remember this is industrial basic research—that is, insurance. Sure, PARC invented the laser printer and the computer network and perfected the graphical user interface and something that came to be known as what-you-see-is-what-you-get computing on a large computer screen, but the captains of industry at Xerox headquarters in Stamford, Connecticut, were making too much money the old way—by making copiers—to remake Xerox into a computer company. They took a couple of halfhearted stabs, introducing systems like the Xerox Star, but generally did little to promote PARC technology. From a business standpoint, Xerox probably did the right thing, but in the long term, failing to develop PARC technology alienated the PARC geniuses.

In his 1921 book *The Engineers and the Price System*, economist

Thorstein Veblen pointed out that in high-tech businesses, the true value of a company is found not in its physical assets but in the minds of its scientists and engineers. No factory could continue to operate if the knowledge of how to design its products and fix its tools of production was lost. Veblen suggested that the engineers simply organize and refuse to work until they were given control of industry. By the 1970s, though, the value of computer companies was so highly concentrated in the programmers and engineers that there was not much to demand control of. It was easier for disgruntled engineers just to walk, taking with them in their minds 70 or 80 percent of what they needed to start a new company. Just add money.

From inside their ivory tower, Taylor's geniuses saw less able engineers and scientists starting companies of their own and getting rich. As it became clear that Xerox was going to do little or nothing with their technology, some of the bolder CSL veterans began to hit the road as entrepreneurs in their own right, founding several of the most important personal computer hardware and software companies of the 1980s. They took with them Xerox technology—its look and feel too. And they took Bob Taylor's model for running a successful high-tech enterprise—a model that turned out not to be so perfect after all.

▸   ▸   ▸   ▸   ▸   ▸   ▸   ▸   ▸   ▸   ▸   ▸

CHAIRMAN BILL LEADS THE HAPPY WORKERS IN SONG

William H. Gates III stood in the checkout line at an all-night convenience store near his home in the Laurelhurst section of Seattle. It was about midnight, and he was holding a carton of butter pecan ice cream. The line inched forward, and eventually it was his turn to pay. He put some money on the counter, along with the ice cream, and then began to search his pockets.

"I've got a 50-cents-off coupon here somewhere," he said, giving up on his pants pockets and moving up to search the pockets of his plaid shirt.

The clerk waited, the ice cream melted, the other customers, standing in line with their root beer Slurpies and six-packs of beer, fumed as Gates searched in vain for the coupon.

"Here," said the next shopper in line, throwing down two quarters.

Gates took the money.

"Pay me back when you earn your first million," the 7-11 philanthropist called as Gates and his ice cream faded into the night.

The shoppers just shook their heads. They all knew it was Bill Gates, who on that night in 1990 was approximately a three billion dollar man.

I figure there's some real information in this story of Bill Gates and the ice cream. He *took* the money. What kind of person is this? What kind of person wouldn't dig out his own 50 cents and pay for the ice cream? A person who didn't have the money? Bill Gates has the money. A starving person? Bill Gates has never starved. Some paranoid schizophrenics would have taken the money (some wouldn't, too), but I've heard no claims that Bill Gates is mentally ill. And a kid might take the money—some bright but poorly socialized kid under, say, the age of 9.

Bingo.

My mother lives in Bentonville, Arkansas, a little town in the northwest part of the state, hard near the four corners of Arkansas, Kansas, Missouri, and Oklahoma. Bentonville holds the headquarters of Wal-Mart stores and is the home of Sam Walton, who founded Wal-Mart. Why we care about this is because Sam Walton is maybe the only person in America who could just write a check and buy out Bill Gates and because my mother keeps running into Sam Walton in the bank.

Sam Walton will act as our control billionaire in this study.

Sam Walton started poor, running a Ben Franklin store in Newport, Arkansas, just after the war. He still drives a pickup truck today and has made his money selling one post hole digger, one fifty-pound bag of dog food, one cheap polyester shirt at a time, but the fact that he's worth billions of dollars still gives him a lot in common with Bill Gates. Both are smart businessmen, both are highly competitive, both dominate their industries, both have been fairly careful with their money. But Sam Walton is old, and Bill Gates is young. Sam Walton has bone cancer and looks a little shorter on each visit to the bank, while Bill Gates is pouring money into biotechnology companies, looking for eternal youth. Sam Walton has promised his fortune to support education in Arkansas, and Bill Gates's representatives tell fund raisers from Seattle charities that their boss is still, "too young to be a pillar of his community."

They're right. He *is* too young.

Our fifteen-minutes-of-fame culture makes us all too quickly pin labels of good or bad on public figures. Books like this one paint their major characters in black or white, and sometimes in red. It's hard to make such generalizations, though, about Bill Gates, who is not really a bad person. In many ways he's not a particularly good person either. What he is is a *young* person, and that was originally by coincidence, but now it's by design. At 36, Gates has gone from being the youngest person to be a self-made billionaire to being the self-made billionaire who acts the youngest.

Spend a late afternoon sitting at any shopping mall. Better still, spend a day at a suburban high school. Watch the white kids and listen to what they say. It's a shallow world they live in—one that's dominated by school and popular culture and by yearning for the opposite sex. Saddam Hussein doesn't matter unless his name is the answer to a question on next period's social studies quiz. Music matters. Clothes matter, unless deliberately stating that they don't matter is part of your particular style. Going to the prom matters. And zits—zits matter *a lot*.

Watch these kids and remember when we were that age and everything was so wonderful and horrible and hormones ruled our lives. It's another culture they live in—another planet even —one that we help them to create. On the white kids' planet, all that is supposed to matter is getting good grades, going to the prom, and getting into the right college. There are no taxes; there is no further responsibility. Steal a car, get caught, and your name doesn't even make it into the newspaper, because you are a juvenile, a citizen of the white kids' planet, where even grand theft auto is a two-dimensional act.

Pay attention now, because here comes the important part.

William H. Gates III, who is not a bad person, is two-dimensional too. Girls, cars, and intense competition in a technology business are his life. Buying shirts, taking regular showers, getting married and being a father, becoming a pillar of his community, and just plain making an effort to get along with other people if he doesn't feel like it are not parts of his life. Those parts belong to someone else—to his adult alter ego. Those parts still belong his father, William H. Gates II.

In the days before Microsoft, back when Gates was a nerdy Harvard freshman and devoting himself to playing high-stakes poker on a more-or-less full-time basis, his nickname was Trey—the gambler's term for a three of any suit. Trey, as in William H. Gates the Trey. His very identity then, as now, was defined in terms of his father. And remember that a trey, while a low card, still beats a deuce.

Young Bill Gates is incredibly competitive because he has a terrific need to win. Give him an advantage, and he'll take it. *Allow* him an advantage, and he'll still take it. Lend him 50 cents and, well, you know . . . . Those who think he cheats to win are generally wrong. What's right is that Gates doesn't mind winning ungracefully. A win is still a win.

It's clear that if Bill Gates thinks he can't win, he won't play. This was true at Harvard, where he considered a career in mathematics until it became clear that there were better undergraduate mathematicians in Cambridge than Bill Gates. And that was true at home in Seattle, where his father, a successful corporate attorney and local big shot, still sets the standard for parenthood, civic responsibility, and adulthood in general.

"There are aspects of his life he's defaulting on, like being a father," said the dad, lobbing a backhand in this battle of generations that will probably be played to the death.

So young Bill, opting out of the adulthood contest for now,

has devoted his life to pressing his every advantage in a business where his father has no presence and no particular experience. That's where the odds are on the son's side and where he's created a supportive environment with other people much like himself, an environment that allows *him* to play the stern daddy role and where he will never ever have to grow old.

Bill Gates's first programming experience came in 1968 at Seattle's posh Lakeside School when the Mothers' Club bought the school access to a time-sharing system. That summer, 12-year-old Bill and his friend Paul Allen, who was two years older, made $4,200 writing an academic scheduling program for the school. An undocumented feature of the program made sure the two boys shared classes with the prettiest girls. Later computing adventures for the two included simulating the northwest power grid for the Bonneville Power Administration, which did not know at the time that it was dealing with teenagers, and developing a traffic logging system for the city of Bellevue, Washington.

"Mom, tell them how it worked before," whined young Bill, seeking his mother's support in front of prospective clients for Traf-O-Data after the program bombed during an early sales demonstration.

By his senior year in high school, Gates was employed full time as a programmer for TRW—the only time he has ever had a boss.

Here's the snapshot view of Bill Gates's private life. He lives in a big house in Laurelhurst, with an even bigger house under construction nearby. The most important woman in his life is his mother, Mary, a gregarious Junior League type who helps run her son's life through yellow Post-it notes left throughout his home.

Like a younger Hugh Hefner, or perhaps like an emperor of China trapped within the Forbidden City, Gates is not even held responsible for his own personal appearance. When Chairman

Bill appears in public with unwashed hair and unkempt clothing, his keepers in Microsoft corporate PR know that they, not Bill, will soon be getting a complaining call from the ever-watchful Mary Gates.

The second most important woman in Bill Gates's life is probably his housekeeper, with whom he communicates mainly through a personal graphical user interface—a large white board that sits in Gates's bedroom. Through check boxes, fill in the blanks, and various icons, Bill can communicate his need for dinner at 8 or for a new pair of socks (brown), all without having to speak or be seen.

Coming from the clothes-are-not-important school of fashion, all of Gates's clothes are purchased by his mother or his housekeeper.

"He really should have his colors done," one of the women of Microsoft said to me as we watched Chairman Bill make a presentation in his favorite tan suit and green tie.

Do us all a favor, Bill; ditch the tan suit.

The third most important woman in Bill Gates's life is the designated girlfriend. She has a name and a face that changes regularly, because nobody can get too close to Bill, who simply will not marry as long as his parents live. No, he didn't say that. I did.

Most of Gates's energy is saved for the Boys' Club—212 acres of forested office park in Redmond, Washington, where 10,000 workers wait to do his bidding. Everything there, too, is Bill-centric, there is little or no adult supervision, and the soft drinks are free.

∽∽

Bill Gates is the Henry Ford of the personal computer industry. He is the father, the grandfather, the uncle, and the godfather of the PC, present at the microcomputer's birth and determined to be there at its end. Just ask him. Bill Gates is the only head honcho I have met in this business who isn't angry, and that's not because he's any weirder than the others—each is weird in his own way—but because he is the only head honcho who is not in a hurry. The others are all trying like hell to get somewhere else before the market changes and their roofs fall in, while Gates is happy right where he is.

Gates and Ford are similar types. Technically gifted, self-centered, and eccentric, they were both slightly ahead of their times and took advantage of that fact. Ford was working on standardization, mass production, and interchangeable parts back when most car buyers were still wealthy enthusiasts, roads were unpaved, and automobiles were generally built by hand. Gates was vowing to put "a computer on every desk and in every home running Microsoft software" when there were fewer than a hundred microcomputers in the world. Each man consciously worked to create an industry out of something that sure looked like a hobby to everyone else.

A list of Ford's competitors from 1908, when he began mass producing cars at the River Rouge plant, would hold very few names that are still in the car business today. Cadillac, Oldsmobile—that's about it. Nearly every other Ford competitor from those days is gone and forgotten. The same can be said for a list of Microsoft competitors from 1975. None of those companies still exists.

Looking through the premier issue of my own rag, *InfoWorld*, I found nineteen advertisers in that 1979 edition, which was then known as the *Intelligent Machines Journal*. Of those nineteen advertisers, seventeen are no longer in business. Other than Microsoft, the only survivor is the MicroDoctor—one guy in Palo Alto who has been repairing computers in the same storefront on El Camino Real since 1978. Believe me, the

MicroDoctor, who at this point describes his career as a preferable alternative to living under a bridge somewhere, has never appeared on anyone's list of Microsoft competitors.

So why are Ford and Microsoft still around when their contemporaries are nearly all gone? Part of the answer has to do with the inevitably high failure rate of companies in new industries; hundreds of small automobile companies were born and died in the first twenty years of this century, and hundreds of small aircraft companies climbed and then power dived in the second twenty years. But an element not to be discounted in this industrial Darwinism is sheer determination. Both Gates and Ford were determined to be long-term factors in their industries. Their objective was to be around fifteen or fifty years later, still calling the shots and running the companies they had started. Most of their competitors just wanted to make money. Both Ford and Gates also worked hard to maintain total control over their operations, which meant waiting as long as possible before selling shares to the public. Ford Motor Co. didn't go public until nearly a decade after Henry Ford's death.

Talk to a hundred personal computer entrepreneurs, and ninety-nine of them won't be able to predict what they will be doing for a living five years from now. This is not because they expect to fail in their current ventures but because they expect to get bored and move on. Nearly every high-tech enterprise is built on the idea of working like crazy for three to five years and then selling out for a vast amount of money. Nobody worries about how the pension plan stacks up because nobody expects to be around to collect a pension. Nobody loses sleep over whether their current business will be a factor in the market ten or twenty years from now—nobody, that is, except Bill Gates, who clearly intends to be as successful in the next century as he is in this one and without having to change jobs to do it.

At 19, Bill Gates saw his life's work laid out before him. Bill, the self-proclaimed god of software, said in 1975 that there will be a Microsoft and that it will exist for all eternity, selling sorta okay software to the masses until the end of time. Actually, the sorta okay part came along later, and I am sure that Bill intended always for Microsoft's products to be the best in their fields. But then Ford intended his cars to be best, but he settled, instead, for just making them the most popular. Gates, too, has had to make some compromises to meet his longevity goals for Microsoft.

Both Ford and Gates surrounded themselves with yes-men and -women, whose allegiance is to the leader rather than to the business. Bad idea. It reached the point at Ford where one suddenly out-of-favor executive learned that he was fired when he found his desk had been hacked to pieces with an ax. It's not like that at Microsoft yet, but emotions do run high, and Chairman Bill is still young.

As Ford did, Gates typically refuses to listen to negative opinions and dismisses negative people from his mind. There is little room for constructive criticism. The need is so great at Microsoft for news to be good that warnings signs are ignored and major problems are often overlooked until it is too late. Planning to enter the PC database market, for example, Microsoft spent millions on a project code-named Omega, which came within a few weeks of shipping in 1990, even though the product didn't come close to doing what it was supposed to do.

The program manager for Omega, who was so intent on successfully bringing together his enormous project, reported only good news to his superiors when, in fact, there were serious problems with the software. It would have been like introducing a new car that didn't have brakes or a reverse gear. Cruising toward a major marketplace embarrassment, Microsoft was saved only through the efforts of brave souls who presented Mike Maples, head of Microsoft's applications division, with a list of promised Omega features that didn't exist. Maples invited the program manager to demonstrate his product, then asked him to demonstrate each of the non-features. The Omega introduction was cancelled that afternoon.

From the beginning, Bill Gates knew that microcomputers would be big business and that it was his destiny to stand at the center of this growing industry. Software, much more than hardware, was the key to making microcomputers a success, and Gates knew it. He imagined that someday there would be millions of computers on desks and in homes, and he saw Microsoft playing *the* central role in making this future a reality. His goal for Microsoft in those days was a simple one: monopoly.

"We want to monopolize the software business," Gates said time and again in the late 1970s. He tried to say it in the 1980s too, but by then Microsoft had public relations people and anti-trust lawyers in place to tell their young leader that the M word was not on the approved corporate vocabulary list. But it's what he meant. Bill Gates had supreme confidence that he knew better than anyone else how software ought to be developed and that his standards would become the de facto standards for the fledgling industry. He could imagine a world in which users would buy personal computers that used Microsoft operating systems, Microsoft languages, and Microsoft applications. In fact, it was difficult, even painful, for Gates to imagine a world organized any other way. He's a very stubborn guy about such things, to the point of annoyance.

The only problem with this grand vision of future computing —with Bill Gates setting all the standards, making all the decisions, and monopolizing all the random-access memory in the world—was that one person alone couldn't do it. He needed help. In the first few years at Microsoft, when the company had fewer than fifty employees and everyone took turns at the switchboard for fifteen minutes each day, Gates could impose his will by reading all the computer code written by the other programmers and making changes. In fact, he rewrote nearly everything, which bugged the hell out of programmers when they had done perfectly fine work only to have it be rewritten (and not necessarily improved) by peripatetic Bill Gates. As Microsoft grew, though, it became obvious that reading every line and rewriting every other wasn't a feasible way to continue. Gates needed to find an instrument, a method of governing his creation.

Henry Ford had been able to rule *his* industrial empire through the instrument of the assembly line. The assembly-line worker was a machine that ate lunch and went home each night to sleep in a bed. On the assembly line, workers had no choice about what they did or how they did it; each acted as a mute extension of Ford's will. No Model T would go out with four headlights instead of two, and none would be painted a color other than black because two headlights and black paint were what Mr. Ford specified for the cars coming off his assembly line. Bill Gates wanted an assembly line, too, but such a thing had never before been applied to the writing of software.

Writing software is just that—writing. And writing doesn't work very well on an assembly line. Novels written by committee are usually not good novels, and computer programs written by large groups usually aren't very good either. Gates wanted to create an enormous enterprise that would supply most of the world's microcomputer software, but to do so he had to find a way to impose his vision, his standards, on what he expected would become thousands of programmers writing millions of lines of code—more than he could ever personally read.

Good programmers don't usually make good business leaders. Programmers are typically introverted, have awkward social skills, and often aren't very good about paying their own bills, much less fighting to close deals and get customers to pay up. This ability to be so good at one thing and so bad at another stems mainly, I think, from the fact that programming is an individual sport, where the best work is done, more often than not, just to prove that it *can* be done rather than to meet any corporate goal.

Each programmer wants to be the best in his crowd, even if that means wanting the others to be not quite so good. This trend, added to the hated burden of meetings and having to care about things like group objectives, morale, and organizational minutiae, can put those bosses who still think of themselves primarily as programmers at odds with the very employees on whom they rely for the overall success of the company. Bill Gates is this way, and his bitter rivalry with nearly every other sentient being on the planet could have been his undoing.

To realize his dream, Gates had to create a corporate structure at Microsoft that would allow him to be both industry titan and top programmer. He had to invent a system that would satisfy his own adolescent need to dominate and his adult need to inspire. How did he do it?

Mind control.

The instrument that allowed Microsoft to grow yet remain under the creative thumb of Bill Gates walked in the door one day in 1979. The instrument's name was Charles Simonyi.

Unlike most American computer nerds, Charles Simonyi was raised in an intellectually supportive environment that encouraged both thinking and expression. The typical American nerd was a smart kid turned inward, concentrating on science and technology because it was more reliable than the world of adult reality. The nerds withdrew into their own society, which logically excluded their parents, except as chauffeurs and financiers. Bill Gates was the son of a big-shot Seattle lawyer who didn't understand his kid. But Charles Simonyi grew up in Hungary during the 1950s, the son of an electrical engineering professor who saw problem solving as an integral part of growing up. And problem solving is what computer programming is all about.

In contrast to the parents of most American computer nerds, who usually had little to offer their too-smart sons and daughters, the elder Simonyi managed to play an important role in his son's intellectual development, qualifying, I suppose, for the Ward Cleaver Award for Quantitative Fathering.

"My father's rule was to imagine that you have the solution already," Simonyi remembered. "It is a great way to solve problems. I'd ask him a question: How many horses does it take to do something? And he'd answer right away, 'Five horses; can you tell me if I am right or wrong?' By the time I'd figured out that it couldn't be five, he'd say, 'Well if it's not five, then it must be X. Can you solve for that?' And I could, because the problem was already laid out from the test of whether five horses was correct. Doing it backward removed the anxiety from the answer. The anxiety, of course, is the fear that the problem can't be solved—at least not by me."

With the help of his father, Simonyi became Hungary's first teenage computer hacker. That's hacker in the old sense of being a good programmer who has a positive emotional relationship with the machine he is programming. The new sense of hacker —the *Time* and *Newsweek* versions of hackers as technopunks and cyberbandits, tromping through computer systems wearing hobnail boots, leaving footprints, or worse, on the innocent data of others—those hackers aren't real hackers at all, at least not to me. Go read another book for stories about those people.

Charles Simonyi was a hacker in the purest sense: he slept with his computer. Simonyi's father helped him get a job as a night watchman when he was 16 years old, guarding the Russian-built Ural II computer at the university. The Ural II had 2,000 vacuum tubes, at least one of which would overheat and burn out each time the computer was turned on. This meant that the first hour of each day was spent finding that burned-out vacuum tube and replacing it. The best way to avoid vacuum tube failure was to leave the computer running all night, so young Simonyi offered to stay up with the computer, guarding and playing with it. Each night, the teenager was in total control of probably half the computing resources in the entire country.

Not that half the computer resources of Hungary were much in today's terms. The Ural II had 4,000 bytes of memory and took eighty microseconds to add two numbers together. This performance and amount of memory was comparable to an early Apple II. Of course the Ural II was somewhat bigger than an Apple II, filling an entire room. And it had a very different user interface; rather than a video terminal or a stack of punch cards, it used an input device much like an old mechanical cash register. The zeroes and ones of binary machine language were punched on cash register–like mechanical buttons and then entered as a line of data by smashing the big ENTER key on the right side. Click-click-click-click-click-click-click-click—SMASH!

Months of smashing that ENTER key during long nights spent learning the innards of the Ural II with its hundreds of blinking lights started Simonyi toward a career in computing. By 1966, he had moved to Denmark and was working as a professional programmer on his first computer with transistors rather than vacuum tubes. The Danish system still had no operating system, though. By 1967, Simonyi was an undergraduate computer science student at the University of California, working on a Control Data supercomputer in Berkeley. Still not yet 20, Simonyi had lived and programmed his way through nearly the entire history of von Neumann–type computing, beginning in the time warp that was Hungary.

By the 1970s, Simonyi was the token skinny Hungarian at Xerox PARC, where his greatest achievement was Bravo, the what-you-see-is-what-you-get word processing software for the Alto workstation.

While PARC was the best place in the world to be doing computer science in those days, its elitism bothered Simonyi, who couldn't seem to (or didn't want to) shake his socialist upbringing. Remember that at PARC there were no junior researchers, because Bob Taylor didn't believe in them. Everyone in Taylor's lab had to be the best in his field so that the Computer Science Lab could continue to produce its miracles of technology while remaining within Taylor's arbitrary limit of fifty staffers. Simonyi wanted larger staffs, including junior people, and he wanted to develop products that might reach market in the programmer's lifetime.

PARC technology was amazing, but its lack of reality was equally amazing. For example, one 1978 project, code-named Adam, was a laser-scanned color copier using very advanced emitter-coupled logic semiconductor technology. The project was technically impossible at the time and is only just becoming possible today, more than twelve years later. Since Moore's Law says that semiconductor density doubles every eighteen months, this means that Adam was undertaken approximately eight generations before it would have been technically viable, which is rather like proposing to invent the airplane in the late sixteenth century. With all the other computer knowledge that needed to be gathered and explored, why anyone would bother with a project like Adam completely escaped Charles Simonyi, who spent lots of time railing against PARC purism and a certain amount of time trying to circumvent it.

This was the case with Bravo. The Alto computer, with its beautiful bit-mapped white-on-black screen, needed software, but there were no extra PARC brains to spare to write programs for it. Money wasn't a problem, but manpower was; it was almost impossible to hire additional people at the Computer

Science Laboratory because of the arduous hiring gauntlet and Taylor's reluctance to manage extra heads. When heads were added, they were nearly always Ph.D.s, and the problem with Ph.D.s is that they are headstrong; they won't do what you tell them to. At least they wouldn't do what Charles Simonyi told them to do. Simonyi did not have a Ph.D.

Simonyi came up with a scam. He proposed a research project to study programmer productivity and how to increase it. In the course of the study, test subjects would be paid to write software under Simonyi's supervision. The test subjects would be Stanford computer science students. The software they would write was Bravo, Simonyi's proposed editor for the Alto. By calling them research subjects rather than programmers, he was able to bring some worker bees into PARC.

The Bravo experiment was a complete success, and the word processing program was one of the first examples of software that presented document images on-screen that were identical to the eventual printed output. Beyond Bravo, the scam even provided data for Simonyi's own dissertation, plunking him right into the ranks of the PARC unmanageable. His 1976 paper was titled "Meta-Programming: A Software Production Method."

Simonyi's dissertation was an attempt to describe a more efficient method of organizing programmers to write software. Since software development will always expand to fill all available time (it does not matter how much time is allotted—software is *never* early), his paper dealt with how to get more work done in the limited time that is typically available. Looking back at his Bravo experience, Simonyi concluded that simply adding more programmers to the team was not the correct method for meeting a rapidly approaching deadline. Adding more programmers just increased the amount of communication overhead needed to keep the many programmers all working in the same direction. This additional overhead was nearly always enough to absorb any extra manpower, so adding more heads to a project just meant that more money was being spent to reach the same objective at the same time as would have the original, smaller, group. The trick to improving programming productivity was making better use of the programmers already in place rather than adding more programmers. Simonyi's method of doing this was to create the position of metaprogrammer.

The metaprogrammer was the designer, decision maker, and communication controller in a software development group. As the metaprogrammer on Bravo, Simonyi mapped out the basic design for the editor, deciding what it would look like to the user and what would be the underlying code structure. But he did not write any actual computer code; Simonyi prepared a document that described Bravo in enough detail that his "research subjects" could write the code that brought each feature to life on-screen.

Once the overall program was designed, the metaprogrammer's job switched to handling communication in the programming group and making decisions. The metaprogrammer was like a general contractor, coordinating all the subcontractor programmers, telling them what to do, evaluating their work in progress, and making any required decisions. Individual programmers were allowed to make no design decisions about the project. All they did was write the code as described by the metaprogrammer, who made all the decisions and made them just as fast as he could, because Simonyi calculated that it was more important for decisions to be made quickly in such a situation than that they be made well. As long as at least 85 percent of the metaprogrammer's interim decisions were ultimately correct (a percentage Simonyi felt confident that he, at least, could reach more or less on the basis of instinct), there was more to be lost than gained by thoughtful deliberation.

The metaprogrammer also coordinated communication

among the individual programmers. Like a telephone operator, the metaprogrammer was at the center of all interprogrammer communication. A programmer with a problem or a question would take it to the metaprogrammer, who could come up with an answer or transfer the question or problem to another programmer who the metaprogrammer felt might have the answer. The alternative was to allow free discussion of the problem, which might involve many programmers working in parallel on the problem, using up too much of the group's time.

By centralizing design, decision making, and communication in a single metaprogrammer, Simonyi felt that software could be developed more efficiently and faster. The key to the plan's success, of course, was finding a class of obedient programmers who would not contest the metaprogrammer's decisions.

The irony in this metaprogrammer concept is that Simonyi, who bitched and moaned so much about the elitism of Xerox PARC, had, in his dissertation, built a vastly more rigid structure that replaced elitism with authoritarianism.

In the fluid structure of Taylor's lab at PARC, only the elite could survive the demanding intellectual environment. In order to bring junior people into the development organization, Simonyi promoted an elite of one—the metaprogrammer. Both Taylor's organization at CSL and Simonyi's metaprogrammer system had hub and spoke structures, though at CSL, most decision making was distributed to the research groups themselves, which is what made it even possible for Simonyi to perpetrate the scam that produced Bravo. In Simonyi's system, only the metaprogrammer had the power to decide.

Simonyi, the Hungarian, instinctively chose to emulate the planned economy of his native country in his idealized software development team. Metaprogramming was collective farming of software. But like collective farming, it didn't work very well.

By 1979, the glamor of Xerox PARC had begun to fade for Simonyi. "For a long while I believed the value we created at PARC was so great, it was worth the losses," he said. "But in fact, the ideas were good, but the work could be recreated. So PARC was not unique.

"They had no sense of business at all. I remember a PARC lunch when a director (this was after the oil shock) argued that oil has no price elasticity. I thought, 'What am I doing working here with this Bozo?' "

Many of the more entrepreneurial PARC techno-gods had already left to start or join other ventures. One of the first to go was Bob Metcalfe, the Ethernet guy, who left to become a consultant and then started his own networking company to exploit the potential of Ethernet that he thought was being ignored by Xerox. Planning his own break for the outside world with its bigger bucks and intellectual homogeneity, Simonyi asked Metcalfe whom he should approach about a job in industry. Metcalfe produced a list of ten names, with Bill Gates at the top. Simonyi never got around to calling the other nine.

When Simonyi moved north from California to join Microsoft in 1979, he brought with him two treasures for Bill Gates. First was his experience in developing software applications. There are four types of software in the microcomputer business: operating systems like Gary Kildall's CP/M, programming languages like Bill Gates's BASIC, applications like VisiCalc, and utilities, which are little programs that add extra functions to the other categories. Gates knew a lot about languages, *thought* he knew a lot about operating systems, had no interest in utilities, but knew very little about applications and admitted it.

The success of VisiCalc, which was just taking off when Simonyi came to Microsoft, showed Gates that application software—spreadsheets, word processors, databases and such—was one of the categories he would have to dominate in order to achieve his lofty goals for Microsoft. And Simonyi, who was seven years older, maybe smarter, and coming straight from PARC—Valhalla itself—brought with him just the expertise that Gates would need to start an applications division at Microsoft. They quickly made a list of products to develop, including a spreadsheet, word processor, database, and a long-since-forgotten car navigation system.

The other treasure that Simonyi brought to Microsoft was his dissertation. Unlike PARC, Microsoft didn't have any Ph.D.s before Simonyi signed on, so Gates did as much research on the Hungarian as he could, which included having a look at the thesis. Reading through the paper, Gates saw in Simonyi's metaprogrammer just the instrument he needed to rule a vastly larger Microsoft with as much authority as he then ruled the company in 1979, when it had around fifty employees.

The term *metaprogrammer* was never used. Gates called it the "software factory," but what he and Simonyi implemented at Microsoft was a hierarchy of metaprogrammers. Unlike Simonyi's original vision, Gates's implementation used several levels of metaprogrammers, which allowed a much larger organization.

Gates was the central metaprogrammer. He made the rules, set the tone, controlled the communications, and made all the technical decisions for the whole company. He surrounded himself with a group of technical leaders called architects. Simonyi was one of these super-nerds, each of whom was given overall responsibility for an area of software development. Each architect was, in turn, a metaprogrammer, surrounded by program managers, the next lower layer of nerd technical management. The programmers who wrote the actual computer code reported to the program managers, who were acting as metaprogrammers, too.

The beauty of the software factory, from Bill Gates's perspective, was that every participant looked strictly toward the center, and at that center stood Chairman Bill—a man so determined to be unique in his own organization that Microsoft had more than 500 employees before hiring its second William.

The irony of all this diabolical plotting and planning is that it did not work. It was clear after less than three months that metaprogramming was a failure. Software development, like the writing of books, is an iterative process. You write a program or a part of a program, and it doesn't work; you improve it, but it still doesn't work very well; you improve it one more time (or twenty more times), and then maybe it ships to users. With all decisions being made at the top and all information supposedly flowing down from the metaprogrammer to the 22-year-old peon programmers, the reverse flow of information required to make the changes needed for each improved iteration wasn't planned for. Either the software was never finished, or it was poorly optimized, as was the case with the Xerox Star, the only computer I know of that had its system software developed in this way. The Star was a dog.

The software factory broke down, and Microsoft quickly went back to writing code the same way everyone else did. But the structure of architects and program managers was left in place, with Bill Gates still more or less controlling it all from the center. And since a control structure was all that Chairman Bill had ever really wanted, he at least considered the software factory to be a success.

Through the architects and program managers, Gates was able to control the work of every programmer at Microsoft, but to do so reliably required cheap and obedient labor. Gates set a policy that consciously avoided hiring experienced programmers, specializing, instead, in recent computer science graduates.

Microsoft became a kind of cult. By hiring inexperienced

workers and indoctrinating them into a religion that taught the concept that metaprogrammers were better than mere programmers and that Bill Gates, as the *meta*metaprogrammer, was perfect, Microsoft created a system of hero worship that extended Gates's will into every aspect of the lives of employees he had not even met. It worked for Kim Il Sung in North Korea, and it works in the suburbs east of Seattle too.

Most software companies hire the friends of current employees, but Microsoft hires kids right out of college and relocates them. The company's appetite for new programming meat is nearly insatiable. One year Microsoft got in trouble with the government of India for hiring nearly every computer science graduate in the country and moving them all to Redmond.

So here are these thousands of neophyte programmers, away from home in their first working situation. All their friends are Microsoft programmers. Bill is a father/folk hero. All they talk about is what Bill said yesterday and what Bill did last week. And since they don't have much to do except talk about Bill and work, there you find them at 2:00 A.M., writing code between hockey matches in the hallway.

Microsoft programmers work incredibly long hours, most of them unproductive. It's like a Japanese company where overtime has a symbolic importance and workers stay late, puttering around the office doing little or nothing just because that's what everyone else does. That's what Chairman Bill does, or is supposed to do, because the troops rarely even see him. *I* probably see more of Bill Gates than entry-level programmers do.

At Microsoft it's a "disadvantage" to be married or "have any other priority but work," according to a middle manager who was unlucky enough to have her secretly taped words later played in court as evidence in a case claiming that Microsoft discriminates against married employees. She described Microsoft as a company where employees were expected to be single or live a "singles lifestyle," and said the company wanted employees that "ate, breathed, slept, and drank Microsoft," and felt it was "the best thing in the world."

The real wonder in this particular episode is not that Microsoft discriminates against married employees, but that the manager involved was a woman. Women have had a hard time working up through the ranks. Only two women have ever made it to the vice-presidential level—Ida Cole and Jean Richardson. Both were hired away from Apple at a time when Microsoft was coming under federal scrutiny for possible sex discrimination. Richardson lasted a few months in Redmond, while Cole stayed until all her stock options vested, though she was eventually demoted from her job as vice-president.

Like any successful cult, sacrifice and penance and the idea that the deity is perfect and his priests are better than you works at Microsoft. Each level, from Gates on down, screams at the next, goading and humiliating them. And while you can work any eighty hours per week that you want, dress any way that you like, you can't talk back in a meeting when your boss says you are shit in front of all your co-workers. It just isn't done. When Bill Gates says that he could do in a weekend what you've failed to do in a week or a month, he's lying, but you don't know better and just go back to try harder.

This all works to the advantage of Gates, who gets away with murder until the kids eventually realize that this is not the way the rest of the world works. But by then it is three or four years later, they've made their contributions to Microsoft, and are ready to be replaced by another group of kids straight out of school.

My secret suspicion is that Microsoft's cult of personality hides a deep-down fear on Gates's part that maybe he doesn't really know it all. A few times I've seen him cornered by some techie who is not from Microsoft and not in awe, a techie who knows more about the subject at hand than Bill Gates ever will. I've seen a flash of fear in Gates's eyes then. Even with you or me, topics can range beyond Bill's grasp, and that's when he uses his "I don't know how technical you are" line. Sometimes this really means that he doesn't want to talk over your head, but just as often it means that he's the one who really doesn't know what he's talking about and is using this put-down as an excuse for changing the subject. To take this particularly degrading weapon out of his hands forever, I propose that should you ever talk with Bill Gates and hear him say, "I don't know how technical you are," reply by saying, that you don't know how technical *he* is. It will drive him nuts.

The software factory allowed Bill Gates to build and control an enormous software development organization that operates as an extension of himself. The system can produce lots of applications, programming languages, and operating systems on a regular basis and at relatively low cost, but there is a price for this success: the loss of genius. The software factory allows for only a single genius—Bill Gates. But since Bill Gates doesn't actually write the code in Microsoft's software, that means that few flashes of genius make their way into the products. They are derivative—successful, but derivative. Gates deals with this problem through a massive force of will, telling himself and the rest of the world that commercial success and technical merit are one in the same. They aren't. He says that Microsoft, which is a superior marketing company, is also a technical innovator. It isn't.

The people of Microsoft, too, choose to believe that their products are state of the art. Not to do so would be to dispute Chairman Bill, which just is not done. It's easier to distort reality.

Charles Simonyi accepts Microsoft mediocrity as an inevitable price paid to create a large organization. "The risk of genius is that the products that result from genius often don't have much to do with each other," he explained. "We are trying to build core technologies that can be used in a lot of products. That is more valuable than genius.

"True geniuses are very valuable if they are motivated. That's how you start a company—around a genius. At our stage of growth, it's not that valuable. The ability to synthesize, organize, and get people to sign off on an idea or project is what we need now, and those are different skills."

Simonyi started Microsoft's applications group in 1979, and the first application was, of course, a spreadsheet. Other applications soon followed as Simonyi and Gates built the development organization they knew would be needed when microcomputing finally hit the big time, and Microsoft would take its position ahead of all its competitors. All they had to do was be ready and wait.

In the software business, as in most manufacturing industries, there are inventive organizations and maintenance organizations. Dan Bricklin, who invented VisiCalc, the first spreadsheet, ran an inventive organization. So did Gary Kildall, who developed CP/M, the first microcomputer operating system. Maintenance organizations are those, like Microsoft, that generally produce derivative products—the second spreadsheet or yet another version of an established programming language. BASIC was, after all, a language that had been placed in the public domain a decade before Bill Gates and Paul Allen decided to write their version for the Altair.

When Gates said, "I want to be the IBM of software," he consciously wanted to be a monolith. But unconsciously he wanted to emulate IBM, which meant having a reactive strategy, multiple divisions, poor internal communications.

As inventive organizations grow and mature, they often

convert themselves into maintenance organizations, dedicated to doing revisions of formerly inventive products and boring as hell for the original programmers who were used to living on adrenalin rushes and junk food. This transition time, from inventive to maintenance, is a time of crisis for these companies and their founders.

Metaprogrammers, and especially nested hierarchies of metaprogrammers, won't function in inventive organizations, where the troops are too irreverent and too smart to be controlled. But metaprogrammers work just fine at Microsoft, which has never been an inventive organization and so has never suffered the crisis that accompanies that fall from grace when the inventive nerds discover that it's only a job.

▸　▸　▸　▸　▸　▸　▸　▸　▸　▸　▸　▸

ALL IBM STORIES
ARE TRUE

I live in California in a house that I can't really afford in a neighborhood filled with blue-haired widows and with two-earner couples who have already made the jump from BMW to Acura and in their hearts are flirting with voting Republican.

Remember when life came mainly in black and white, and Wally and the Beav walked down a street as the credits rolled across them? That was *my house* they walked by on that tree-lined street, my 50-by-105 foot lot, my gnawing termites, my 1957 Studebaker Golden Hawk dripping oil in the driveway, and my orange tree dropping oranges in the side yard. For a kid raised in Apple Creek, Ohio, walking out the door in the middle of winter and pulling a fresh orange off your own tree is heaven.

And New York City is hell.

The New York I reluctantly visit is filled with angry people, garbage, burned coffee, potholes, and overcooked vegetables. Yet I have lots of friends who live in Manhattan and tell me it's the most wonderful place. *Their* New York has theater, music, museums, a great library, and Central Park, while all I have at my house in Palo Alto is good phone service, one orange tree, and a bookshelf containing the complete works of Louis L'Amour. My friends are certain they would die of boredom in Palo Alto, while I'm just as sure I'd die in New York, though not of boredom.

Who's right? We both are.

There's a psychological principle called cognitive dissonance at work here. When the reward (having an orange tree) is out of proportion to the effort required to achieve it (a $3,000 monthly mortgage payment), I am put in a state of dissonance, which can be resolved only by selling the house and moving back to Apple Creek or by warping my entire sense of values to convince myself that it's worth all the effort.

Of course, I choose to warp reality. We all do. I tell myself that warm winters and fresh oranges are worth *anything*, while my friends in New York say exactly the same thing about plays that they don't really attend and parks they are afraid to visit.

Cognitive dissonance plays a major role in all of our lives, and so far it's the only reason I can come up with that most people continue to work for IBM.

Here's what I mean. In early 1983, a big guy named Don Estridge was about to become IBM's vice-president of worldwide manufacturing, making about $250,000 per year in salary and bonuses. What does a vice-president of worldwide manufacturing do, you ask? Not much in a company like IBM, where each division does its own manufacturing while fighting with all the others. Estridge's VP role was really a holding pattern designed to stash a guy who had served the company well but was now out of sync with the Brooks Brothers reality of what the largest computer company in the world was all about.

Since 1980, Estridge had been head of the team that invented the IBM PC, moving Big Blue in ways and at a speed that it had never been moved before. Estridge had been told to go out and break the rules, did as he was told, and by doing so became a dangerous man. Once his team had succeeded by specifically being as *unlike* the rest of IBM as possible, Estridge could no longer be trusted because his ways were no longer the ways of IBM. Welcome to the vice-presidency of worldwide manufacturing.

Then one day in early 1983, 28-year-old Steve Jobs showed up on Don Estridge's doorstep in Boca Raton, Florida, looking for a new president for Apple Computer. Jobs has only three ways of dealing with people: he seduces them, castigates them, or ignores them. In fact, everyone in Jobs's life eventually runs through all three modes, sometimes more than once. Jobs was in seduction mode when he approached Estridge, and Steve can be incredibly seductive when it suits his needs. He is the best salesman in the world and put all those talents to work trying to recruit the tall man from IBM.

Estridge, who had for years programmed his own Apple II at home, took a trip out to Apple headquarters in California and liked what he saw. The money wasn't bad either: $1 million annual salary, $1 million signing bonus, and a $2 million loan to buy that Silicon Valley dream house.

Estridge was about to be stashed in a corporate backwater and would probably never be considered for the top job at IBM. The fit with Apple was nearly perfect, and the money was terrific. Nevertheless, he agonized for a while, talked it over with his closest friends, and turned it down: cognitive dissonance.

Don Estridge couldn't have left IBM even for the top job at Apple. That's because Apple is only a company, while IBM is a *country*.

With annual sales around $60 billion, IBM has a greater gross national product than most countries. It has a relatively stable population of around 380,000 workers. Throw in the spouses and their 1.8 kids each, and we're looking at more than a million citizens of IBM.

Demographically, IBM is most like Kuwait, but temperamentally IBM is Switzerland. Like Switzerland, IBM is conservative, a little dull, slow to change, yet prosperous. Both countries are in the habit of taking in more money than they give out. Both countries learn slowly and adapt at their own pace. Switzerland and IBM can survive anything, or at least think they can. They may be slow, but don't mess with them, because they will fight to keep what is theirs. And if pushed, they'll fight dirty.

Like Switzerland, IBM is landlocked, though Big Blue's barriers are regulation and internal rivalries, not geography. Instead of facing Austria, France, Germany, Italy, and Liechtenstein on its borders, IBM is surrounded by U.S. antitrust laws and a 1956 consent decree that somewhat limits its ability to wreak havoc upon the land. Even more limiting is the rivalry between IBM's different computer divisions, each protecting its turf from incursions by the others. IBM's mainframe division worries as much about competition from the top end of the company's own minicomputer line as it does from any outside competitor. And there is no law or consent decree limiting the amount of infighting that can go on within the company.

The citizens of IBM didn't invent the computer. They don't make the most powerful computers either. The citizens of IBM just make more computers than anybody else. So just as Levis define blue jeans to a world that somehow survived Gloria Vanderbilt, IBM defines computers.

IBM computers don't stand apart, but IBM people do. Of all the companies I've dealt with, the only two whose people consistently present a common front, a kind of unique company style, are those from IBM and Procter & Gamble. This comes from their hiring practices, the way they indoctrinate their workers, and the fact that both companies have had official songbooks. There must be something very unifying about getting together with a thousand other folks at a sales meeting in

New York or Cincinnati and singing your guts out in praise of the Old Man.

The men and women of IBM have their own language and stick to it. A minicomputer is a *mid-range system*. A monitor is a *display*. A hard disk drive—a data storage device that has one or several magnetic platters and spins continuously at 3,600 rpm— is for some reason called a *fixed disk*, although it isn't fixed at all. Sticking to these terms preserves the illusion that IBM's $800 display is somehow different from Samsung's $349 monitor or that IBM's fixed disk drive, made under contract by Seagate, is somehow superior to the exact same drive sold for half price under the Seagate brand.

Like Rolex or Gucci, the people of IBM know that they are not really selling computers at all but the IBM name.

IBM people are a little smug, a little slow, and slightly over-weight. Most IBMers are hired straight out of school and have never worked for another company. They are folks who drive Buick Regals and take them to the car wash every Saturday morning, paying extra to get the hot wax. Their contented mid-dle-class style bugs the hell out of Silicon Valley entrepreneur types, who want to do business with IBM and yet can't under-stand that there are folks in the world—even in the world of computers—who aren't, like them, madly driven to have a for-tune and a Ferrari before their midlife crisis.

IBMers aren't in the business to become millionaires. How can they? These people are not sitting on stock options in some start-up, waiting for their penny shares to go public at $8. They work for a company that went public more than sixty years ago —the quintessential blue chip. Even IBM salespeople, who work on commission, selling computers that can cost millions of dol-lars, have carefully set quotas that effectively limit their earning potential.

Ross Perot, founder of Electronic Data Systems, was one

IBM salesman who got fed up and left the company when he filled his sales quota for the entire year before the end of January and knew that he wouldn't be allowed to sell any more computers—or earn any more money—for eleven more months.

The people of IBM don't need to be rich. They either want the security of working for a company that will employ them for life, offering fringe benefits beyond those of any welfare state, or they want the sheer power that comes from eventually working up into the stratospheric reaches of the most powerful company on earth. Money and power are not synonymous at IBM, where power is preferred.

The price of both prosperity and power is compliance with the rules and the pace of IBM. The rules say that you go where the company asks you to go, do what the company asks you to do, and don't talk about work with strangers. There is a class of company that won't tolerate different behavior, and those companies sometimes suffer for it. IBM is like that. The pace is slow because it takes time to get 760,000 legs marching together.

Every IBM employee's ambition is apparently to become a manager, and the company helps them out in this area by making management the company's single biggest business. IBM executives don't design products and write software; they *manage* the design and writing of software. They go to meetings. So much effort, in fact, is put into managing all the managers who are managing things that hardly anyone is left over to do the real work. This means that most IBM hardware and nearly all IBM software is written or designed by the lowest level of people in the company—trainees. Everyone else is too busy going to meetings, managing, or learning to be a manager, so there is little chance to include any of their technical expertise in IBM products.

Go back and read that last paragraph over again, because that's why IBM products often aren't very competitive.

IBM has layers and layers of management to check and

verify each decision as it is made and amended. The safety net is so big at IBM that it is hard to make a bad decision. In fact, it is hard to make any decision at all, which turns out to be the company's greatest problem and the source of its ultimate downfall (remember, you read it here first).

Except at the very highest positions in IBM, this corporate support system produces a class of executives with bovine, cud-chewing dispositions, who think only on command and typically rely on the company to tell them what to do and when to do it. Before beginning each new assignment, for example, IBM people are thoroughly briefed with all the information the company believes they will need to know in order to do their new job. The briefings are so complete that most IBM people don't bother to do any outside reading or research of their own. If IBM marketing executives know how their personal computers compare with the competition's, it is nearly always through their briefing books and hardly ever by actually using the other guy's hardware—or even their own.

And at the top of IBM, where synapses do pop on occasion, and brain activity is usually, though not always, measurable, nearly all of that activity goes into playing corporate political games, as though competitors and even the global computer market didn't exist.

It was corporate infighting, in fact, that finally made entry into the microcomputer market so attractive to IBMers grown tired of slugging out the next point of mainframe market share while at the same time engaging in internecine warfare with other company divisions. In the microcomputer business, there looked to be no divisional rivalries to worry about, no antitrust considerations, and, most important, the customers were all new, fresh meat, having never before felt the firm handshake of an IBM salesrep. Every sales dollar brought in to buy a microcomputer would be a dollar that would not otherwise have come to IBM. There was something pure about that, and the IBM executives who led the company's assault on the microcomputer market knew that success on this new battlefield could eventually lead them to the real font of power: IBM worldwide headquarters in Armonk, New York.

There were lots of players in the microcomputer business back in 1980, but Apple, Atari, Commodore, and Radio Shack all looked about the same to IBM—small. Total U.S. microcomputer sales had reached $1 billion, so there was a market worth dominating. And as Apple had in 1977, IBM saw the market being sales to small businesses, a segment that the company had previously touched only through its typewriter operation.

The IBM Personal Computer that eventually came to market in late 1981 came from a renegade independent business unit based in Boca Raton, Florida. This wasn't IBM's first try at developing a microcomputer. At least four other designs had been proposed to management in Armonk, including one earlier design from Boca. The major difference between the project that eventually produced the IBM PC and these earlier efforts was that the group of men brought together in July 1980 by Entry Systems Division (ESD) lab director Bill Lowe were pledged to do their work in real time, not IBM time. They had just one year to bring their product to market.

A year is no time at all to IBM. At the time that Lowe got the go-ahead for Project Chess, which would produce Acorn, code name for the IBM PC, there was an ESD project called Datamaster entering its fourth year of development with no end in sight. Datamaster was an attempt to add some microcomputer functions to IBM's Displaywriter dedicated word-processing system. If it took IBM four years just to throw additional functions into an existing product, building an entire computer system in one year must have looked impossible.

It *was* impossible, and Lowe knew it. There was no way that IBM could develop a personal computer in a year. The best they could do was to gather hardware and software from other companies, get them to work together as a system, and then slap an IBM label on the outside. That's what Bill Lowe decided to do.

The question was whether to build or buy. Up to this point, whenever IBM had been faced with the choice between building a component in an IBM plant or buying it from an outside supplier, the decision was always build, build, build. This was because power within Big Blue was measured in part by the number of workers under each manager. Workers at some supplier's plant in Bayonne didn't count.

But Lowe and his crew, breaking their first of many rules, decided to buy everything. They started by looking for software. Since Lowe wanted to buy his operating system software from an established vendor, CP/M looked like his only choice. CP/M came from Gary Kildall's Digital Research, only for some reason IBM didn't know that. The usually infallible briefing book said that CP/M was a Microsoft product. In probably his last gracious gesture toward a competitor, Bill Gates told the caller from IBM that a mistake had been made and gave them Kildall's number in Pacific Grove.

There was still room for IBM and Microsoft to do business, though, because the other software component that seemed to be required in a 1980 microcomputer was a built-in version of the BASIC programming language. Apple, Commodore, and Radio Shack all came with built-in BASIC so users could write their own simple programs. If Acorn was to compete successfully against these machines, IBM would need a BASIC, too. Microsoft was the oldest and best-known provider of microcomputer BASIC and was IBM's probable vendor choice.

It didn't hurt, either, that Mary Gates sat on the national board of United Way along with IBM chairman John Opel and that the two had become friends. Opel was impressed that Lowe was talking with Microsoft and said so, cinching the deal.

Proposing to build a new line of computers around the products of a couple of five-year-old software companies from the West Coast was a bold move for Lowe and a risky one. Jump, for a moment, into Bill Lowe's shoes. It's July 1980. The all-powerful IBM Corporate Management Committee (CMC) has just heard his bold proposal to enter the personal computer business within a year. They tell him to come back in a month with details. The whole plan depends on getting reliable suppliers, so Lowe sends his lieutenants out to Digital Research and Microsoft to find out what kind of people these are. When the IBMers arrived in Pacific Grove, California, to talk with Gary Kildall at Digital Research, he wasn't there. Despite his appointment with IBM, Gary had gone flying in his small plane. Not a good first impression.

With Gary out flying around, the people left in charge at Digital Research didn't know what these IBM guys wanted to talk about, and the IBM guys wouldn't talk about anything until a nondisclosure agreement was signed.

Remember that IBM operates at all times under the consent decree of 1956, which limits its ability to compete. Remember, too, that IBM has the largest legal staff of any U.S. corporation, devoted primarily to finding ways to turn what ought to be limitations into advantages. Enter the IBM nondisclosure agreement, which is the legal equivalent of a neutron bomb, destroying only the people but leaving their technology intact.

Nondisclosure agreements place limits on the ability of parties to reveal the secrets of organizations with which they are doing business. IBM's standard nondisclosure agreement goes even further. By signing the IBM agreement, would-be suppliers agree that whatever they tell IBM is not confidential, while whatever IBM tells them *is* confidential. In other words, while the IBM guy on the other side of the table can tell anyone at all what you reveal to him about your company and its plans, he can take you to court if you repeat one of his jokes, much less reveal any IBM secrets. And if IBM takes legal action, the agreement prohibits the other party from even offering a defense.

IBM's Big Brotherish explanation of its nondisclosure agreement is that would-be suppliers can protect their secrets simply by not revealing them to IBM representatives. In fact, the agreement urges signers not to reveal anything that is confidential. But how can companies do business without revealing confidential information? They can't. Little companies that want to do business with IBM must sign the we-win, you-lose nondisclosure agreement, bare their corporate souls, and pray that an eventual IBM contract makes it all worthwhile. Big companies that talk with IBM throw their own nondisclosure agreement on the table and demand that IBM sign it too.

Jump back to Pacific Grove, where Digital Research didn't even have a nondisclosure agreement of its own. Gary was still flying around somewhere over the Santa Cruz mountains, while Dorothy Kildall squinted at the IBM nondisclosure agreement, imagining her new house with its stable and hot tub going on the auction block following an IBM legal action. She refused to sign, so the men from IBM left town, having never revealed their plans for Acorn but still needing an operating system.

The IBMers headed north for their meeting at Microsoft headquarters in Bellevue, Washington. At this time, Microsoft had about fifty employees and was selling versions of the FORTRAN and COBOL computer languages in addition to its many varieties of BASIC. Microsoft had a hardware division, too, that produced a circuit card that made it possible for Apple II computers to run the CP/M operating system, which at that time had better word processing and database programs than were available on the Apple. Microsoft's Softcard, and the operating system software that shipped with it, made Bill Gates at that time the largest seller of

Gary Kildall's CP/M. Gates knew what a good business selling operating systems could be.

Gates, Paul Allen, and Gates's buddy from Harvard, Steve Ballmer, put on neckties for a change and met with the IBM crowd. They shrugged and signed the nondisclosure agreement, enjoyed a short period of bewildering small talk, gave a tour of the building, and saw the IBM contingent to the door. Nobody from IBM mentioned that the company was working on a personal computer; that came in the next meeting a few days later.

At their second meeting, IBM asked Microsoft to supply a BASIC language for its new computer. According to Gates, the design they described to him was for an 8-bit computer similar to any of the many CP/M systems already on the market. Gates urged the IBM group to go instead with one of the new 16-bit processors just being released. Using a 16-bit processor would make the IBM PC seem more powerful than its competitors, and it would allow the machine to use more memory too. Apple IIs at that time were limited to around 48,000 characters, or "bytes" of memory—usually called 48K. Most CP/M machines hit the wall at 64K. A 16-bit processor could address vastly more memory than competing machines, offering a marketing advantage and a clear growth path for the future.

The argument was persuasive, but IBM still didn't have an 8-bit operating system, much less one that would run on 16-bit machines. Gates said he might be able to help out with an operating system.

Choosing a 16-bit processor was easy. Intel, Motorola, and National Semiconductor were all shipping 16-bit processors at the time. Intel had the 8086 and 8088 processors, Motorola had the 68000, and National had its 16032. The National processor was elegant and powerful; the Motorola was powerful and easy to write software for; the Intel 8086 was fairly powerful but had an awkward memory architecture; the Intel 8088 was an 8086 without the power.

Of course, IBM chose the 8088—the least attractive of all the processors from a technical standpoint. In this case, technical considerations took a back seat to IBM's manufacturing and marketing concerns. The plan was to build a computer without any custom components—just off-the-shelf parts from major semiconductor makers. The 8088 was the only 16-bit processor for which there was available a full complement of the support chips required to build a computer. Motorola and National were still working on their 16-bit support chips, as was Intel for the 8086. But the 8088 was a 16-bit processor in an 8-bit body, since it used an 8-bit data bus—sending and receiving data 8 bits at a time and then processing them in 16-bit mode. This 8-bit bus is what made the 8088 less powerful than the other contenders, but it also made it possible for the 8088 to use support chips intended for the earlier 8080 family of Intel 8-bit processors. Since the 8088 was the only processor that could be used without developing custom support chips, it was the only processor that fit IBM's needs.

The 8088 was cheaper than the other processors because it could use the older support chips and because Intel had deliberately priced it below the 8086. But price was not a major consideration in IBM's decision.

The fact that the 8088 wasn't as powerful as the other processors was actually seen by IBM as an advantage, since it meant that Acorn would not draw customer attention from the company's mid-range systems, which were only slightly faster in many applications, though they cost thousands more. Although Project Chess was going ahead without input from IBM's other divisions, Lowe knew better than to ask for trouble, so the embryonic PC was made deliberately slower than it might have been. It would have been fairly easy too to give Acorn the ability to emulate any or all of IBM's terminals so the new PC could be used to communicate with IBM minicomputers and mainframes. But that sales advantage was deliberately avoided because it would have killed the company's terminal business.

But wait. What about the operating system?

At the moment IBM was having its second meeting with Bill Gates, there were no 16-bit microcomputer operating systems on the market. If IBM had waited for Gary Kildall to get back from the airport, it might have learned that Digital Research was already working on CP/M-86, which would run on Intel's 16-bit 8086 microprocessor and on its little brother, the 8088. CP/M-86 would be ready to go about the time that IBM planned to release its personal computer, too, so it would have been a logical choice, had IBM known that CP/M-86 existed. As the largest seller of CP/M software, Microsoft knew that Digital Research had CP/M-86 coming down the chute, yet Gates, Allen, and Ballmer never mentioned it in their second meeting with IBM. Remember that the IBM nondisclosure agreement specifically urged them not to reveal any confidential information. CP/M-86 was clearly confidential.

Gary Kildall thought that he and Gates had divided the software market, with Digital Research taking the operating system business and Microsoft controlling the programming languages. Bill Gates knew better.

Across Lake Washington, at a company called Seattle Computer Products, was the operating system that Bill Gates wanted to sell to IBM. All he had to do was get it.

In business, as in comedy, timing is everything. There was nothing about QDOS, Seattle Computer Products' 16-bit operating system, that couldn't have been created just as well by programmers at Microsoft. But Microsoft programmers hadn't created it, and Tim Paterson of Seattle Computer Products had.

QDOS, which stood for "quick and dirty operating system," was a 16-bit clone of CP/M intended for an 8086-based computer being developed by the small company. All QDOS commands were the same as in CP/M. Paterson admitted to a little "low-level borrowing" from CP/M, too, but claimed that most of the code was his own.

Gary Kildall still thinks a lot of the QDOS code was stolen straight from his CP/M. "Ask Bill why function code 6 [in QDOS and still in MS-DOS, more than ten years later] ends in a dollar sign. No one in the world knows that but me."

There was nothing earthshaking about QDOS, except that it already existed. Bill Gates was buying time more than anything else when he paid Seattle Computer Products $50,000 for rights to the operating system. It must have seemed like a lot of money at the time.

Here's a great scene that never happened. Bill Gates flies to Florida, makes his pitch to IBM, offering to sell it a product called Quick & Dirty DOS, that, by the way, has at least some code stolen line for line from CP/M. The ears of Justice Department lawyers 900 miles away would have perked up. The IBM legal department, which was then suing Fujitsu for stealing IBM code, would have had a corporate seizure. And young Bill Gates would have found himself standing in the sun-drenched IBM parking lot wondering if it was something he said.

Instead, when Gates made that flight to Florida, he kept to the letter of IBM's own nondisclosure agreement and didn't reveal much about the true heritage of QDOS, now called MS-DOS, other than that it had been developed with the help of Seattle Computer Products.

Not everyone at Microsoft was so certain that the company ought to get into the operating system business. Microsoft was already running at full capacity just doing languages, so adding QDOS would require a major expansion. What if Microsoft expanded and then IBM canceled Project Chess at the last moment? Big Blue had already canceled four other microcomputer projects. If IBM canceled the deal and Microsoft couldn't find other customers for QDOS, then that $50,000 paid to Seattle Computer Products really would have been a lot of money. So while Bill Gates has a well deserved reputation for being cheap, his caution in acquiring QDOS was not unfounded. By jumping into bed with IBM, Gates was putting his entire company at risk, and he knew it.

In short order, Microsoft and IBM concluded a co-development agreement making Microsoft responsible for all the system software for Acorn. Gates, Allen, and company would finish the development of QDOS, which would be sold by IBM under the name PC-DOS. They would also provide a BASIC interpreter that would be shipped with each machine in read-only memory. Never before had IBM allowed itself to be so dependent on a single supplier, much less one run by a 25-year-old who ought to wash his hair more often. Aligning with Microsoft was a daring move for Lowe's crew from Boca Raton and a clear indication of how independent they really were from the old way of doing things at IBM. The Microsoft connection made the IBM PC possible, and the IBM connection ensured Microsoft's long-term success as a software company.

Bill Lowe came up with the idea of IBM's building a personal computer. He created a renegade design group, mapped out a design for the original IBM PC, found Microsoft to provide the system software, and sold IBM's Corporate Management Committee on the project. Lowe's reward, which came in late 1980, was a promotion to vice-president in another division of IBM.

Lowe's bosses at IBM saw building the organization, rather than the PC, as the object of their enterprise. It did not matter, then, that promoting Lowe meant taking him away from his brainchild; the company was built on large teams and the idea that no one person could be critical to an IBM venture. And Lowe, who had worked at IBM for eighteen years, toed the organization line by gratefully accepting the promotion to run IBM's lab in Rochester, Minnesota.

In most other PC companies, the person in Lowe's position would have been kept on the project until it was complete. Any change—even accompanied by a promotion—would have been viewed as a sign of disfavor. But Bill Lowe was a company man and went where the company said he should go.

Lowe's hand-picked successor was Don Estridge, who carried out Lowe's vision for Acorn. Most of this vision was defined not by how the computer would be but how it would *not* be. For one thing, it would not be sold by only IBM's direct sales force because there was no way that a company geared to direct sales of million-dollar mainframes could use the same sales force to sell microcomputers profitably to end users at the prevailing $3,000 price point. Most Acorns would be sold in retail stores. The first computer retailers to carry IBM products were the ComputerLand chain and Sears Business Centers—Sears Roebuck & Co.'s plunge into the office equipment business that would coincide with the IBM PC's introduction.

IBM was deliberately bringing a higher, and more boring, level of professionalism to the business of selling computers. You can imagine that the first visit of Sears representatives to Boca was very different from their first visit to Atari's computer plant in Sunnyvale, California. When the Sears buyers came to the Atari plant, they found a typical Silicon Valley tilt-up building filled with noisy production machinery, rock music, and pot smoke. Atari founder Nolan Bushnell gave the Sears crowd a tour by sitting them in cardboard cartons and literally running them down the assembly line.

IBM did things differently.

In order to meet its price point, profit margins, and delivery schedule, a lot of features were left out of the basic PC. The base machine, for example, would ship without serial or parallel ports for adding telephone modems or printers. No video graphics capability was built in. And while QDOS could theoretically address up to 640K of memory, the basic machine came with only 16K, which could be expanded to a maximum of only 64K by switching to higher-capacity memory chips on the main circuit board, called the motherboard. It was possible to add serial and parallel ports, graphics capability, and memory beyond 64K to the PC, but that required adding extra circuit cards that fit in special slots on the motherboard.

The Apple II had used circuit cards to add memory and features because such cards were the norm in HP 3000 minicomputers used by Steve Wozniak in his earlier job designing hand-held calculators at Hewlett-Packard. Woz designed his own data bus, or scheme for adding special function cards to the Apple II. Most CP/M machines used another bus, called the S-100, for adding special features and memory. An Apple II card would not work in an S-100 computer, nor would an S-100 card fit in an Apple II. There were many manufacturers of Apple II and S-100 cards because both bus standards were published by their inventors, with full information available about how to design a card to work in each type of computer. IBM engineers came up with yet another bus design for the PC, and like the other companies, they published their bus specification so that third parties could design cards to go in IBM PCs.

IBM's rationale for publishing its bus specification was a good one. For one thing, the consent decree of 1956 required that they publish the technical specifications of all their products, though that was one rule that had occasionally been overlooked in the past. More important, with only a year to build Acorn, there just wasn't enough time to develop many add-in circuit cards, so IBM would do serial and parallel cards and a video card, but most other such cards would have to be left to third parties to develop. Also, IBM was going to have a hard enough time making acceptable profit margins on Acorn itself, and an analysis of the add-on card business looked even worse, so the decision was made to leave those crumbs to outsiders. Other companies would have to assume the development and marketing costs of add-on cards, but the existence of such cards would only help sales of IBM PCs.

Had anyone in Boca bothered to notice, Acorn was not going to be a very proprietary machine. Microsoft retained the rights to sell QDOS to companies other than IBM. Every component in the new machine came straight from some semiconductor company's stock bins, with the exception of the ROM-BIOS chip, which linked IBM's hardware to Microsoft's operating system software. The bus specification was published and available to any manufacturer who wanted to implement it. All of this meant that there was not much keeping other computer companies from building computers exactly like the IBM PC, piggybacking on Big Blue's probable success in the microcomputer market. All that a would-be clone maker would have to do is reverse engineer the ROM-BIOS, something that Amdahl Corp. was already doing in the IBM mainframe world and that the courts had decided was legal. But IBM was not worried about others' copying its microcomputer because to do so would require buying the same chips from the same suppliers as IBM, though probably in lower volumes and hence at higher prices than Big Blue was paying. The planners in Boca saw how other companies could copy Acorn, but they did not see how it could be done at a profit.

The only thing remarkable about the IBM Personal Computer was that it was designed and built by IBM. The PC was deliberately positioned against the Apple II. Compared to the Apple II, the PC was big and clunky, but that was by design. It was supposed to look more like a piece of office equipment, while the Apple II was at home on a shelf underneath the family television. The PC had a bigger screen, a larger keyboard on a long cord, and floppy disk drives that held more data (160K each) than did Apple II floppies. Not that anyone could imagine needing 640K of random access memory—that's how much you could pack into an IBM PC, that is, after some third-party manufacturer came up with a circuit card that held enough memory chips.

The PC looked substantial and had slots, lots of slots for add-in cards. Here was a machine that looked as if it could be expanded forever. Never mind that the anemic power supply was not strong enough to power a PC with all slots filled.

The PC had its husky look, its new operating system (though one that looked reassuringly like CP/M). It had the IBM name and a massive promotion budget to go with it. The new machine even had applications that were provided by third parties but initially released under an IBM label. There was the EasyWriter word processing package and two spreadsheets—Multiplan, from Microsoft, and VisiCalc.

The PC was a big success and rapidly became the top-selling microcomputer. But it wasn't a significantly better VisiCalc machine than was the Apple II or its follow-on, the Apple III. VisiCalc and Multiplan looked exactly the same on the screen of an Apple II or an IBM PC, and neither program was significantly faster on the IBM platform either. If it was going to realize its full potential, the PC would still need a compelling application—one that offered features never before seen on a personal computer and that tied those features specifically to the IBM PC platform so buyers would see no choice but to buy an IBM PC. Every successful computer needs at least one of these compelling applications, remember?

▸   ▸   ▸   ▸   ▸   ▸   ▸   ▸   ▸   ▸   ▸   ▸

SOFTWARE ENVY

Mitch Kapor, the father of Lotus 1-2-3, showed up one day at my house but wouldn't come inside. "You have a cat in there, don't you?" he asked.

Not one cat but two, I confessed. I am a sinner.

Mitch is allergic to cats. I mean *really* allergic, with an industrial-strength asthmatic reaction. "It's only happened a couple of times," he explained, "but both times I thought I was going to die."

People have said they are dying to see me, but Kapor really means it.

At this point we were still standing in the front yard, next to Kapor's blue rental car. The guy had just flown cross-country in a Canadair Challenger business jet that costs $3,000 per hour to run, and he was driving a $28.95-per-day compact from Avis. I would have at least popped for a T-Bird.

We were still standing in the front yard because Mitch Kapor needed to use the bathroom, and his mind was churning out a risk/reward calculation, deciding whether to chance contact with the fierce Lisa and Jeri, our kitty sisters.

"They are generally sleeping on the clean laundry about this time," I assured him.

He decided to take a chance and go for it.

"You won't regret it," I called after him.

Actually, I think Mitch Kapor has quite a few regrets. Success has placed a heavy burden on Mitch Kapor.

Mitch is a guy who was in the right place at the right time and saw clearly what had to be done to get very, very rich in record time. Sure enough, the Brooklyn-born former grad student, recreational drug user, disc jockey, Transcendental Meditation teacher, mental ward counselor, and so-so computer programmer today has a $6 million house on 22 acres in Brookline, Massachusetts, the $12 million jet, and probably the world's foremost collection of vintage Hawaiian shirts. So why isn't he happy?

I think Mitch Kapor isn't happy because he feels like an imposter.

This imposter thing is a big problem for America, with effects that go far beyond Mitch Kapor. Imposters are people who feel that they haven't earned their success, haven't paid their dues—that it was all too easy. It isn't enough to be smart, we're taught. We have to be smart, *and* hard working, *and* long suffering. We're supposed to be aggressive and successful, but our success is not supposed to come at the expense of anyone else. Impossible, right?

We got away from this idea for a while in the 1980s, when Michael Milken and Donald Trump made it okay to be successful on brains and balls alone, but look what's happened to *them*. The tide has turned against the easy bucks, even if those bucks are the product of high intelligence craftily applied, as in the case of Kapor and most of the other computer millionaires. We're in a resurgence of what I call the guilt system, which can be traced back through our educational institutions all the way to the medieval guild system.

The guild system, with its apprentices, journeymen, and masters, was designed from the start to screen out people, not encourage them. It took six years of apprenticeship to become a journeyman blacksmith. Should it really take *six years* for a reasonably intelligent person to learn how to forge iron? Of course not. The long apprenticeship period was designed to keep newcomers out of the trade while at the same time rewarding those at the top of the profession by giving them a stream of young helpers who worked practically for free.

This concept of dues paying and restraint of trade continues in our education system today, where the route to a degree is typically cluttered with requirements and restrictions that have little or nothing to do with what it was we came to study. We grant instant celebrity to the New Kids on the Block but support an educational system that takes an average of eight years to issue each Ph.D.

The trick is to not put up with the bullshit of the guild system. That's what Bill Gates did, or he would have stayed at Harvard and become a near-great mathematician. That's what Kapor did, too, in coming up with 1-2-3, but now he's lost his nerve and is paying an emotional price. Doe-eyed Mitch Kapor has scruples, and he's needlessly suffering for them.

We're all imposters in a way—I sure am—but poor Mitch feels guilty about it. He knows that it's not brilliance, just cleverness, that's the foundation of his fortune. What's wrong with that? He knows that timing and good luck played a much larger part in the success of 1-2-3 than did technical innovation. He knows that without Dan Bricklin and VisiCalc, 1-2-3 and the Kapor house and the Kapor jet and the Kapor shirt collection would never have happened.

"Relax and enjoy it," I say, but Mitch Kapor won't relax. Instead, he crisscrosses the country in his jet, trying to convince himself and the world that 1-2-3 was not a fluke and that he can do it all again. He's also trying to convince universities that they ought to promote a new career path called software designer, which is the name he has devised for his proto-technical function. A software designer is a smart person who thinks a lot about software but isn't a very good programmer. If Kapor is successful in this educational campaign, his career path will be legitimized and be made guilt free but at the cost of others having to pay dues, not knowing that they shouldn't really have to.

꒰꒱

"Good artists copy," said Pablo Picasso. "*Great* artists steal."

I like this quotation for a lot of reasons, but mainly I like it because the person who told it to me was Steve Jobs, co-founder of Apple Computer, virtual inventor of the personal computer business as it exists today, and a died-in-the-wool sociopath. Sometimes it takes a guy like Steve to tell things like they really are.

And the way things really are in the computer business is that there is a whole lot of copying going on. The truly great ideas are sucked up quickly by competitors, and then spit back on the market in new products that are basically the old products with slight variations added to improve performance and keep within the bounds of legality. Sometimes the difference between one computer or software program and the next seems like the difference between positions 63 and 64 in the *Kama Sutra*, where 64 is the same as 63 but with pinkies extended.

The reason for this copying is that there just aren't very many really great ideas in the computer business—ideas good enough and sweeping enough to build entire new market segments around. Large or small, computers all work pretty much the same way—not much room for earth-shaking changes there. On the software side, there are programs that simulate physical systems, or programs that manipulate numbers (spreadsheets), text and graphics (word processors and drawing programs), or raw data (databases). And that's about the extent of our genius so far in horizontal applications—programs expected to appeal to nearly every computer user.

These apparent limits on the range of creativity mean that Dan Bricklin invented the first spreadsheet, but you and I didn't, and we never can. Despite our massive intelligence and good looks, the best that we can hope to do is invent the *next* spreadsheet or maybe the *best* spreadsheet, at least until our product, too, is surpassed. With rare exceptions, what computer software and hardware engineers are doing every day is reinventing things. Reinventing isn't easy, either, but it can still be very profitable.

The key to profitable reinvention lies in understanding the relationship between computer hardware and software. We know that computers have to exist before programmers will write software specifically for them. We also know that people usually buy computers to run a single compelling software application. Now we add in longevity—the fact that computers die young but software lives on, nearly forever. It's always been this way. Books crumble over time, but the words contained in those books—the software—survive as long as readers are still buying and publishers are still printing new editions. Computers don't crumble—in fact, they don't even wear out—but the physical boxes are made obsolete by newer generations of hardware long before the programs and data inside have lost their value.

What software does lose in the transition from one hardware generation to the next is an intimate relationship with that hardware. Writing VisiCalc for the Apple II, Bob Frankston had the Apple hardware clearly in mind at all times and optimized his work to run on that machine by writing in assembly language—the internal language of the Apple II's MOStek 6502 microprocessor—rather than in some higher-level language like BASIC or

FORTRAN. When VisiCalc was later translated to run on other types of computers, it lost some of that early intimacy, and performance suffered.

But even if intimacy is lost, software hangs on because it is so hard to produce and so expensive to change.

Moore's Law says that the number of transistors that can be built on a given area of silicon doubles every eighteen months, which means that a new generation of faster computer hardware appears every eighteen months too. Cringely's Law (I just thought this up) says that people who actually rely on computers in their work won't tolerate being more than one hardware generation behind the leading edge. So everyone who can afford to buys a new computer when their present computer is three years old. But do all these users get totally new software every time they buy a new computer to run it on? Not usually, because the training costs of learning to use a new application are often higher than the cost of the new computer to run it on.

Once the accounting firm Ernst & Young, with its 30,000 personal computers, standardizes on an application, it takes an act of God or the IRS to change software.

Software is more complex than hardware, though most of us don't see it that way. It seems as if it should be harder to build computers, with their hundreds or thousands of electrical connections, than to write software, where it's a matter of just saying to the program that a connection exists, right? But that isn't so. After all, it's easier to print books than it is to write them.

Try typing on a computer keyboard. What's happening in there that makes the letters appear on the screen? Type the words "Cringely's mom wears army boots" while running a spreadsheet program, then using a word processor, then a different word processor, then a database. The internal workings of each program will handle the words differently—sometimes radically differently—from the others, yet all run on the same hardware and all yield the same army boots.

Woz designed and built the Apple I all by himself in a couple of months of spare time. Even the prototype IBM PC was slapped together by half a dozen engineers in less than thirty days. Software is harder because it takes the hardware only as a starting point and can branch off in one or many directions, each involving levels of complexity far beyond that of the original machine that just happens to hold the program. Computers are house scaled, while software is building scaled.

The more complex an application is, the longer it will stay in use. It shouldn't be that way, but it is. By the time a program grows to a million lines of code, it's too complex to change because no one person can understand it all. That's why there are mainframe computer programs still running that are more than 30 years old.

In software, there are lots of different ways of solving the same problem. VisiCalc, the original spreadsheet, came up with the idea of cells that had row and column addresses. Right from the start, the screen was filled with these empty cells, and without the cells and their addresses, no work could be done. The second spreadsheet program to come along was called T/Maker and was written by Peter Roizen. T/Maker did not use cells at all and started with a blank screen. If you wanted to total three rows of numbers in T/Maker, you put three plus signs down the left-hand side of the screen as you entered the numbers and then put an equal sign at the bottom to indicate that was the place to show a total. T/Maker also included the ability to put blocks of text in the spreadsheet, and it could even run text vertically as well as horizontally. VisiCalc had nothing like that.

A later spreadsheet, called Framework and written by Robert Carr, replaced cells with what Carr called frames. There were different kinds of frames in Framework, with different properties—

like row-oriented frames and column-oriented frames, for example. Put some row-oriented frames inside a single column-oriented frame, and you had a spreadsheet. That spreadsheet could then be put as a nested layer inside another spreadsheet also built of frames. Mix and match your frames differently, and you had a database or a word processor, all without a cell in sight.

If VisiCalc was an apple, then T/Maker was an orange, and Framework was a rutabaga, yet all three programs could run on identical hardware, and all could produce similar output although through very different means. *That's* what I mean by software being more complex than hardware.

Having gone through the agony of developing an application or operating system, then, software developers have a great incentive to greet the next generation of hardware by translating the present software—"porting" it—to the new environment rather than starting over and developing a whole new version that takes complete advantage of the new hardware features.

It's at this intersection of old software and new hardware that the opportunity exists for new applications to take command of the market, offering extra features, combined with higher performance made possible by the fact that the new program was written from scratch for the new computer. This is one of the reasons that WordStar, which once ruled the market for CP/M word processing programs, is only a minor player in today's MS-DOS world, eclipsed by WordPerfect, a word processing package that was originally designed to run on Data General minicomputers but was completely rewritten for the IBM PC platform.

In both hardware and software, successful reinvention takes place along the edges of established markets. It's usually not enough just to make another computer or program like all the others; the new product has to be superior in at least one respect. Reinvented products have to be cheaper, or more powerful, or smaller, or have more features than the more established products with which they are intended to compete. These are all examples of edges. Offer a product that is in no way cheaper, faster, or more versatile—that skirts no edges—and buyers will see no reason to switch from the current best-seller.

Even the IBM PC skirted the edges by offering both a 16-bit processor and the IBM nameplate, which were two clear points of differentiation.

Once IBM's Personal Computer was established as the top-selling microcomputer in America, it not only followed a market edge, it created one. Small, quick-moving companies saw that they had a few months to make enduring places for themselves purely by being the first to build hardware and software add-ons for the IBM PC. The most ambitious of these companies bet their futures on IBM's success. A hardware company from Cleveland called Tecmar Inc. camped staffers overnight on the doorstep of the Sears Business Center in Chicago to buy the first two IBM PCs ever sold. Within hours, the two PCs were back in Ohio, yielding up their technical secrets to Tecmar's logic analyzers.

And on the software side, Lotus Development Corp. in Cambridge, Massachusetts, bet nearly $4 million on IBM and on the idea that Lotus 1-2-3 would become the compelling application that would sell the new PC. A spreadsheet program, 1-2-3 became the single most successful computer application of all.

Mitch Kapor had a vision, a moment of astounding insight when it became obvious to him how and why he should write a spreadsheet program like 1-2-3. *Vision* is a popular word in the computer business and one that has never been fully defined—until now. Just what the heck does it mean to have such a vision?

George Bush called it the "vision thing." Vision—high-tech executives seem to bathe in it or at least want us to think that they do. They are "technical visionaries," having their "technical visions" so often, and with such blinding insight, that it's probably not safe for them to drive by themselves on the freeway. The truth is that technical vision is not such a big deal.

Dan Bricklin's figuring out the spreadsheet, *that's* a big deal, but it doesn't fit the usual definition of technical vision, which is the ability to foresee potential in the work of others. Sure, some engineer working in the bowels of IBM may think he's come up with something terrific, but it takes having his boss's boss's boss's boss think so, too, and say so at some industry pow-wow before we're into the territory of vision. Dan Bricklin's inventing the spreadsheet was a bloody miracle, but Mitch Kapor's squinting at the IBM PC and figuring out that it would soon be the dominant microcomputer hardware platform—that's *vision*.

There, the secret's out: vision is only seeing neat stuff and recognizing its market potential. It's reading in the newspaper that a new highway is going to be built and then quickly putting up a gas station or a fast food joint on what is now a stretch of country road but will soon be a freeway exit.

Most of the so-called visionaries don't program and don't design computers—or at least they haven't done so for many years. The advantages these people have are that they are listened to by others and, because they are listened to by others, all the *real* technical people who want the world to know about the neat stuff they are working on seek out these visionaries and give them demonstrations. Potential visions are popping out at these folks all the time. All they have to do is sort through the visions and apply some common sense.

Common sense told Mitch Kapor that IBM would succeed in the personal computer business but that even IBM would require a compelling application—a spreadsheet written from scratch to take advantage of the PC platform—to take off in the market. Kapor, who had a pretty fair idea of what was coming down the tube from most of the major software companies, was amazed that nobody seemed to be working on such a native-mode PC spreadsheet, leaving the field clear for him. Deciding to do 1-2-3 was a "no brainer."

When IBM introduced its computer, there were already two spreadsheet programs that could run on it—VisiCalc and Multiplan—both ported from other platforms. Either program could have been the compelling application that IBM's Don Estridge knew he would need to make the PC successful. But neither VisiCalc nor Multiplan had the performance, the oomph, required to kick IBM PC sales into second gear, though Estridge didn't know that.

The PC sure looked successful. In the four months that it was available at the end of 1981, IBM sold about 50,000 personal computers, while Apple sold only 135,000 computers for the entire calendar year. By early 1982, the PC was outselling Apple two-to-one, primarily by attracting first-time buyers who were impressed by the IBM name rather than by a compelling application.

At the end of 1981, there were 2 million microcomputers in America. Today there are more than 45 million IBM-compatible PCs alone, with another 10 million to 12 million sold each year. It's this latter level of success, where sales of 50,000 units would go almost unnoticed, that requires a compelling application. That application—Lotus 1-2-3—didn't appear until January 26, 1983.

Dan Bricklin made a big mistake when he didn't try to get a patent on the spreadsheet. After several software patent cases had gone unsuccessfully as far as the U.S. Supreme Court, the general thinking when VisiCalc appeared in 1979 was that software could not be patented, only copyrighted. Like the words of a book, the individual characters of code could be protected by a copyright, and even the specific commands could be protected, but what couldn't be protected by a copyright was the literal function performed by the program. There is no way that a copyright could protect the idea of a spreadsheet. Protecting the idea would have required a patent.

Ideas are strange stuff. Sure, you could draw up a better mousetrap and get a patent on that, as long as the Patent Office saw the trap design as "new, useful, and unobvious." A spreadsheet, though, had no physical manifestation other than a particular rhythm of flashing electrons inside a microprocessor. It was that specific rhythm, rather than the actual spreadsheet function it performed, that could be covered by a copyright. Where the patent law seemed to give way was in its apparent failure to accept the idea of a spreadsheet as a virtual machine. VisiCalc was performing work there in the computer, just as a mechanical machine would. It was doing things that could have been accomplished, though far more laboriously, by cams, gears, and sprockets.

In fact, had Dan Bricklin drawn up an idea for a mechanical spreadsheet machine, it would have been patentable, and the patent would have protected not only that particular use for gears and sprockets but also the underlying idea of the spreadsheet. Such a patent would have even protected that idea as it might later be implemented in a computer program. That's not what Dan Bricklin did, of course, because he was told that software couldn't be patented. So he got a copyright instead, and the difference to Bricklin between one piece of legal paper and the other was only a matter of several hundred million dollars.

On May 26, 1981, after seven years of legal struggle, S. Pal Asija, a programmer and patent lawyer, received the first software patent for SwiftAnswer, a data retrieval program that was never heard from again and whose only historical function was to prove that all of the experts were wrong; software *could* be patented. Asija showed that when the Supreme Court had ruled against previous software patent efforts, it wasn't saying that software was unpatentable but that those particular programs weren't patentable. By then it was too late for Dan Bricklin.

By the time VisiCalc appeared for the IBM PC, Bricklin and Frankston's spreadsheet was already available for most of the top-selling microcomputers. The IBM PC version of VisiCalc was, in fact, a port of a port, having been translated from a version for the Radio Shack TRS-80 computer, which had been translated originally from the Apple II. VisiCalc was already two years old and a little tired. Here was the IBM PC, with up to 640K of memory available to hold programs and extra features, yet still VisiCalc ran in 64K, with the same old feature set you could get on an Apple II or on a "Trash-80." It was no longer compelling to the new users coming into the market. They wanted something new.

Part of the reason VisiCalc was available on so many microcomputers was that Dan Fylstra's company, which had been called Personal Software but by this time was called VisiCorp, wanted out of its contract with Dan Bricklin's company, Software Arts. VisiCorp had outgrown Fylstra's back bedroom in Massachusetts and was ensconced in fancier digs out in California, where the action was. But in the midst of all that Silicon Valley action, VisiCorp was hemorrhaging under its deal with Software Arts, which still paid Bricklin and Frankston a 37.5 percent royalty on each copy of VisiCalc sold. VisiCalc sales at one point reached a peak of 30,000 copies per month, and the agreement required VisiCorp to pay Software Arts nearly $12 million in 1983 alone—far more than either side had ever expected.

Fylstra wanted a new deal that would cost his company less, but he had little power to force a change. A deal was a deal, and hackers like Bricklin and Frankston, whose professional lives were based on understanding and following the strict rules of programming, were not inclined to give up their advantage cheaply. The only coercion entitled VisiCorp under the contract, in fact, was its right to demand that Software Arts port VisiCalc to as many different computers as Fylstra liked. So Fylstra made Bricklin port VisiCalc to every microcomputer.

It was clear to both VisiCorp and Software Arts that the 37.5 percent royalty was too high. Today the usual royalty is around 15 percent. Fylstra wanted to own VisiCalc outright, but in two years of negotiations, the two sides never came to terms.

VisiCorp had published other products under the same onerous royalty schedule. One of those products was VisiPlot/VisiTrend, written by Mitch Kapor and Eric Rosenfield. VisiPlot/VisiTrend was an add-on to VisiCalc; it could import data from VisiCalc and other programs and then plot the data on graphs and apply statistical tests to determine trends from the data. It was a good program for stock market analysis.

VisiPlot/VisiTrend was derived from an earlier Kapor program written during one of his many stints of graduate work, this time at the Sloan School of Management at MIT. Kapor's friend Rosenfield was doing his thesis in statistics using an econometric modeling language called TROLL. To help Rosenfield cut his bill for time on the MIT computer system, Kapor wrote a program he called Tiny TROLL, a microcomputer subset of TROLL. Tiny TROLL was later rewritten to read VisiCalc files, which turned the program into VisiPlot/VisiTrend.

VisiCorp, despite its excessive royalty schedule, was still the most successful microcomputer software company of its time. For its most successful companies, the software business is a license to print money. After the costs of writing applications are covered, profit margins run around 90 percent. VisiPlot/VisiTrend, for example, was a $249.95 product, which was sold to distributors for 60 percent off, or $99.98. Kapor's royalty was 37.5 percent of that, or $37.49 per copy. VisiCorp kept $62.49, out of which the company paid for manufacturing the floppy disks and manuals (probably around $15) and marketing (perhaps $25), still leaving a profit of $22.49. Kapor and Rosenfield earned about $500,000 in royalties for VisiPlot/VisiTrend in 1981 and 1982, which was a lot of money for a product originally intended to save money on the Sloan School time-sharing system but less than a tenth of what Dan Bricklin and Bob Frankston were earning for VisiCalc, VisiCorp's real cash cow. This earnings disparity was not lost on Mitch Kapor.

Kapor learned the software business at VisiCorp. He moved to California for five months to work for Fylstra as a product manager, helping to select and market new products. He saw what was both good and bad about the company and also saw the money that could be made with a compelling application like VisiCalc.

VisiCalc wasn't the only program that VisiCorp wanted to buy outright in order to get out from under that 37.5 percent royalty. In 1982, Roy Folke, who worked for Fylstra, asked Kapor what it would take to buy VisiPlot/VisiTrend. Kapor first asked for $1 million—that magic number in the minds of most programmers, since it's what they always seem to ask for. Then Kapor thought again, realizing that there were other mouths to feed from this sale, other programmers who had helped write the code and deserved to be compensated. The final price was $1.2 million, which sent Mitch Kapor home to Massachusetts with $600,000 after taxes. Only three years before, he had been living in a room in Marv Goldschmitt's house, wondering what to do with his life, and playing with an Apple II he'd hocked his stereo to buy.

Kapor saw the prototype IBM PC when he was working at VisiCorp. He had a sense that the PC and its PC-DOS operating system would set new standards, creating new edges of opportunity. Back in Boston, he took half his money—$300,000—and bet it on this one-two punch of the IBM PC and PC-DOS. It was a gutsy move at the time because experts were divided about the prospects for success of both products. Some pundits saw real benefits to PC-DOS but nothing very special about IBM's hardware.

Others thought IBM hardware would be successful, though probably with a more established operating system. Even IBM was hedging its bets by arranging for two other operating systems to support the PC—CP/M-86 and the UCSD p-System. But the only operating system that shipped at the same time as the PC, and the only operating system that had IBM's name on it, was PC-DOS. That wasn't lost on Mitch Kapor either.

When riding the edges of technology, there is always a question of how close to the edge to be. By choosing to support only the IBM PC under PC-DOS, Kapor was riding damned close to the edge. If both the computer and its operating system took off, Kapor would be rich beyond anyone's dreams. If either product failed to become a standard, 1-2-3 would fail; half his fortune and two years of Kapor's life would have been wasted. Trying to minimize this same risk, other companies adopted more conservative paths. In San Diego, Context Management Systems, for example, was planning an integrated application far more ambitious than Lotus 1-2-3, but just in case IBM and PC-DOS didn't make it, Context MBA was written under the UCSD p-System.

That lowercase *p* stands for *pseudo*. Developed at the University of California at San Diego, the p-System was an operating system intended to work on a wide variety of microprocessors by creating a pseudomachine inside the computer. Rather than writing a program to run on a specific computer like an IBM PC, the idea was to write for this pseudocomputer that existed only in computer memory and ran identically in a number of different computers. The pseudomachine had the same user interface and command set on every computer, whether it was a PC or even a mainframe. While the user programmed the pseudomachine, the pseudomachine programmed the underlying hardware. At least that was the idea.

The p-System gave the same look and feel to several otherwise dissimilar computers, though at the expense of the added pseudomachine translation layer, which made the p-System S-L-O-W—slow but safe, to the minds of the programmers writing Context MBA, who were convinced that portability would give them a competitive edge. It didn't.

Context MBA had a giant spreadsheet, far more powerful than VisiCalc. The program also offered data management operations, graphics, and word processing, all within the big spreadsheet. Like Mitch Kapor and Lotus, Context had hopes for success beyond that of mere mortals.

Context MBA appeared six months before 1-2-3 and had more features than the Lotus product. For a while, this worried Kapor and his new partner, Jonathan Sachs, who even made some changes in 1-2-3 after looking at a copy of Context MBA. But their worries were unfounded because the painfully slow performance of Context MBA, with its extended spreadsheet metaphor and p-System overhead, killed both the product and the company. Lotus 1-2-3, on the other hand, was written from the start as a high-performance program optimized strictly for the IBM PC environment.

Sachs was the programmer for 1-2-3, while Kapor called himself the software designer. A software designer in the Mitch Kapor mold is someone who wears Hawaiian shirts and is intensely interested in the details of a program but not necessarily in the underlying algorithms or code. Kapor stopped being a programmer shortly after the time of Tiny TROLL. The roles of Kapor and Sachs in the development of 1-2-3 generally paralleled those of Dan Bricklin and Bob Frankston in the development of VisiCalc.

The basis of 1-2-3 was a spreadsheet program for Data General minicomputers already written by Sachs, who had worked at Data General and before that at MIT. Kapor wanted to offer several functions in one program to make 1-2-3 stand out from its competitors, so they came up with the idea of adding graphics and a word processor to Sachs's original spreadsheet. This way users could crunch their financial data, prepare graphs and diagrams illustrating the results, and package it all in a report prepared with the word processor. It was the word processor, which was being written by a third programmer, that became a bottleneck, holding up the whole project. Then Sachs played with an early copy of Context MBA and discovered that the word processing module of that product was responsible for much of its poor performance, so they decided to drop the word processor module in 1-2-3 and replace it with a simple database manager, which Sachs wrote, retaining the three modules needed to still call it 1-2-3, as planned.

Unlike Context MBA, Lotus 1-2-3 was written entirely in 8088 assembly language, which made it very fast. The program beat the shit out of Multiplan and VisiCalc when it appeared. (Bill Gates, ever unrealistic when it came to assessing the performance of his own products, predicted that Microsoft's Multiplan would be the death of 1-2-3.) The Lotus product worked only on the PC platform, taking advantage of every part of the hardware. And though the first IBM PCs came with only 16K of onboard memory, 1-2-3 required 256K to run—more than any other microcomputer program up to that time.

Given that Sachs was writing nearly all the 1-2-3 code under the nagging of Kapor, there has to be some question about where all the money was going. Beyond his own $300,000 investment, Kapor collected more than $3 million in venture capital —nearly ten times the amount it took to bring the Apple II computer to market.

The money went mainly for creating an organization to sell 1-2-3 and for rolling out the product. Even in 1983, there were thousands of microcomputer software products vying for shelf space in computer stores. Kapor and a team of consultants from McKinsey & Co. decided to avoid competitors entirely by selling

1-2-3 directly to large corporations. They ignored computer stores and computer publications, advertising instead in *Time* and *Newsweek*. They spent more than $1 million on mass market advertising for the January 1983 roll-out. Their bold objective was to sell up to $4 million worth of 1-2-3 in the first year. As the sellers of a financial planning package, it must have been embarrassing when they outstripped that first-year goal by 1,700 percent. In the first three months that 1-2-3 was on the market, IBM PC sales tripled. Big Blue had found its compelling application, and Mitch Kapor had found his gold mine.

Lotus sold $53 million worth of 1-2-3 in its first year. By 1984, the company had $157 million in sales and 700 employees. One of the McKinsey consultants, Jim Manzi, took over from Kapor that year as president, developing Lotus even further into a marketing-driven company centered around a sales force four times the size of Microsoft's, selling direct to Fortune 1000 companies.

As Lotus grew and the thrill of the start-up turned into the drill of a major corporation, Kapor's interests began to drift. To avoid the imposter label, Kapor felt that he had to follow spectacular success with spectacular success. If 1-2-3 was a big hit, just think how big the next product would be, and the next. A second product was brought out, Symphony, which added word processing and communications functions to 1-2-3. Despite $8 million in roll-out advertising, Symphony was not as big a success as 1-2-3. This had as much to do with the program's "everything but the kitchen sink" total of 600 commands as it did with the $695 price. After Symphony, Lotus introduced Jazz, an integrated package for the Apple Macintosh that was a clear market failure. Lotus was still dependent on 1-2-3 for 80 percent of its royalties and Kapor was losing confidence.

Microsoft made a bid to buy Lotus in 1984. Bill Gates wanted that direct sales force, he wanted 1-2-3, and he wanted once again to be head of the largest microcomputer software company, since the spectacular growth of Lotus had stolen that distinction from Microsoft. Kapor would become Microsoft's third-largest stockholder.

"He seemed happy," said Jon Shirley, who was then president of Microsoft. "We would have made him a ceremonial vice-chairman. Manzi was the one who didn't like the plan."

A merger agreement was reached in principle and then canceled when Manzi, who could see no role for himself in the technically oriented and strong-willed hierarchy of Microsoft, talked Kapor out of it.

Meanwhile, Software Arts and VisiCorp had beaten each other to a pulp in a flurry of lawsuits and countersuits. Meeting by accident on a flight to Atlanta in the spring of 1985, Kapor and Dan Bricklin made a deal to sell Software Arts to Lotus, after which VisiCalc was quickly put to death. Now there was no first spreadsheet, only the best one.

Four senior executives left Lotus in 1985, driven out by Manzi and his need to rebuild Lotus in his own image.

"I'm the nicest person I know," said Manzi.

Then, in July 1986, finding that it was no longer easy and no longer fun, Mitch Kapor resigned suddenly as chairman of Lotus, the company that VisiCalc built.

CLONES

It was in the clay room, a closet filled with plastic bags of gray muck at the back of Mr. Ziska's art room, where I made my move. For the first time ever, I found myself standing alone with Nancy Wilkins, the love of my life, the girl of my dreams. She was a vision in her green and black plaid skirt and white blouse, with little flecks of clay dusted across her glasses. Her blonde hair was in a ponytail, her teeth were in braces, and I was sure—well, pretty sure—that she was wearing a bra.

"Run away with me, Nancy," I said, wrapping my arms around her from behind. Forget for a moment, as I obviously did, that we were both 13 years old, trapped in the eighth grade, and had nowhere to run away to.

"Why would I want to run away?" Nancy responded, gently twisting free. "Let's stay here and have fun with everyone else."

It wasn't a rejection, really. There had been no screams, no slaps, no frenzied pounding on the door by Earl Ziska, eager to throw his 120 pounds of fighting fury against me for making a pass at one of his art students. And she'd used the word *let's,* so maybe I had a chance. Still, Nancy's was a call to mediocrity, to being just like all the other kids.

Running away still sounded better to me.

What I really had in mind was not running away but running toward something, toward a future where I was older (16 would do it, I reckoned) and taller and had lots of money and could live out my fantasies with impunity, Nancy Wilkins at my side. But I couldn't say that. It wouldn't have been cool to say, "Come with me to a place where I am taller."

We never ran anywhere together, Nancy and I. It was clear from that moment in the clay room that she was content to live her life in formation with everyone else's and to limit her goals to within one standard deviation on the upside of average. Like nearly everyone else in school and in the world, she wanted more than anything else to be just like her best friends. Only prettier, of course.

Fitting in is the root of culture. Staying here and having fun with everyone else is what allows societies to function, but it's not a source of progress. Progress comes from discord—from doing new things in new ways, from running away to something new, even when it means giving up that chance to have fun with the old gang.

To engineers—really good ones, interested in making progress—the best of all possible worlds would be one in which technologies competed continuously and only the best technologies survived. Whether the good stuff came from an established company, a start-up, or even from Earl Ziska wouldn't matter. But it usually does matter because the real world, the one we live in, is a world of dollars, not sense. It's a world where commercial interests are entrenched and consumers typically pay closer attention to what everyone else is buying than to whether what they are buying is any good. In this real world, then, the most successful products become standards against which all other products are measured, not for their performance or cleverness but for the extent to which they are like that standard.

In the standards game, as in penmanship, the best grades often go to the least interesting people.

In 1948, CBS introduced the long-playing record album—the LP. The new records spun at 33⅓ revolutions per minute rather than the 78 RPM that had been the standard for forty years. This slower speed, combined with the fact that the smaller needle allowed the grooves to be closer together than on the old 78s, made it possible to put more music than ever before on each side of a record. The sound quality of the LPs was better, too. They called it "stereo high fidelity."

The smaller needle used to play an LP and its light tracking weight meant that records wouldn't wear out as quickly as they had with the old steel needles. And the light needles meant that LPs could be made out of unbreakable vinyl rather than the thick, brittle plastic that had been used before.

LPs were better in every way than the old 78s they replaced. Sure, listeners would have to buy new record players, and LPs might cost more to buy, but those were minor penalties for the glories of true high fidelity.

Also in 1948, at about the same time that CBS was introducing the LP, RCA was across town bringing out the first 45 RPM single. The 45 had a better sound than the old 78s, too, though not as good as the LP and not in stereo. But where the LPs put twenty minutes of music on one record side, the 45s opted for a minimalist solution—one song per side—which made 45s cheaper than the 78s they replaced, and lots cheaper than LPs. Forty-fives worked well in jukeboxes, too, because their large center holes made life easier for robot fingers.

The 45s were pretty terrific, though you still had to buy a new record player.

So here it was 1948. One war was over, and the next one was not even imagined, America and American tastes ruled the world, and the record industry had just offered up its two best ideas for how music should be sold for the next forty years. What happened?

The recording industry immediately entered a four-year slump as Americans, who couldn't decide what type of record to buy, decided not to buy any records at all.

What happened to the record industry in 1948 was the result of two major players' deciding to promote new technical standards at exactly the same time.

"You'll sell millions of 45s," the RCA salesmen told record store owners.

"Just *listen* to the music," said the CBS salesman.

"Who's going to pay six bucks for one record?" asked RCA.

"Think profit margins," ordered CBS.

"Think sales volume!"

Who could think? So they didn't, and the industry fumbled along until an act of God or Elvis Presley decided which standard would dominate what parts of the business. Forty-fives eventually gained the youth vote, while LPs took the high end of the market. In time, machines were built that could play both types of records, and the two technical standards were eventually marketed in a manner that made them complementary. But that wasn't the original intention of their inventors, each of whom wanted to have it all.

Markets hate equality. That was the problem with this battle between LPs and 45s: both were better than the old standard, and each had advantages over the other. In the world of music, circa 1948, it just wasn't immediately clear which standard would be dominant, so the third parties in the industry did not know how to align themselves. If either CBS or RCA had been a couple of years later, the market would have had a chance to adopt the first new standard and then consider the second. Everybody would have been listening to more music.

In any major market, there are always two standards, and generally only two, because people are different enough that they won't all be satisfied with the same thing, yet consumers naturally align themselves into either the "us" or "them" camp. No triangles. Even the Big Three U.S. automakers don't constitute a triangle because they have all chosen to support the same standard—the passenger automobile. For all the high school bickering I remember about whether a Ford was better than a Chevy, the alternative standard to a Mustang is not a Camaro; it's a pickup truck.

Just as there are always two standards, one of those standards is always dominant. Eighty-five percent of the folks who go shopping for a passenger vehicle come home with a car, while 15 percent come home with a truck. Eighty-five percent of the home videocassette recorders in America are VHS, while 15 percent are Betamax. Those numbers—85 and 15—keep coming back again and again. Maybe that's the natural relationship between primary and secondary standards, somehow determined by the gods of consumer products.

In the personal computer business today, about 85 percent of the machines sold are IBM compatible, and 15 percent are Apple Macintoshes. Sure, there are other brands—Commodore Amigas, Atari STs, and weird boxes built in England that function in ways that make sense only to English minds—and even the makers of these machines complain that somehow they have trouble getting noticed by anything but the hobbyist market. The mainstream American market—the business market—just doesn't see these machines as computers, even though some of them offer superior features. It's not that they aren't good; it's that they are third.

When IBM introduced its Personal Computer, the world was ready for a change. The 8-bit computers of the time were doing their best to imitate the battle between LPs and 45s. There just wasn't much of a qualitative difference between the Apple IIs, TRS-80s, and CP/M boxes of the time, so no one standard had broken out, taking the overall market to new heights with it. The market needed differentiation, and that was provided by the entry of IBM, raising its 16-bit standard.

Eight-bit partisans looked down their noses at the new PC, said that it was overpriced and underpowered, and asked who would ever need that much memory, anyway. With 3,000 Apple II applications and 5,000 CP/M applications on the market, sheer volume of software would keep IBM and PC-DOS from succeeding, they argued. Their letters of protest in *InfoWorld* had a note of shrillness, though, as if the writers were suddenly and for the first time aware of their own mortality. That's the way it is with soon-to-be passing standards. Collectors of 78s sounded that way too until they vanished.

In the world of standards, ubiquity is the last step before invisibility.

The new standard was going to be 16-bit computing, that was clear, but what wasn't immediately clear was that the new standard would be 16-bit computing using IBM hardware and the PC-DOS operating system. Many companies saw as much opportunity to build the new 16-bit standard computing with *their* hardware and *their* operating system as with IBM's.

There were lots of IBM competitors. There was the Victor 9000, sold by Kidde, an old-line office machine company. The Victor had more power, more storage, more memory, and better graphics than the IBM PC, and for less money. There was the Zenith Z-100, which had *two* processors, so it could run 8-bit or 16-bit software, and it too was a little cheaper than the IBM PC. There was the Hewlett-Packard HP-150, which had more power, more storage, more memory than the IBM PC, *and* a nifty touch-screen that let users make choices by pointing at the screen.

CLONES

There was the DEC Rainbow 100, which had more power, more
storage, and the DEC name. There was a Xerox computer, a
Wang computer, and a Honeywell computer. There were suddenly lots of 16-bit computers hoping to snatch the mantle of de
facto industry standard away from IBM, through either superior
technology or pricing.

One reason that all these players were trying to take on IBM
was that Microsoft encouraged them to. Bill Gates, too, was un-
certain that IBM's PC-DOS would become the new standard, so
he urged all the other companies doing 16-bit computers with
Intel processors to implement their own versions of DOS. And it
was good business, too, since Microsoft programmers were doing
the work of making MS-DOS work on each new platform. No
matter which company set the standard, Microsoft was determined that it would involve a version of their operating system.

But there was another reason for Microsoft to encourage
IBM's competitors to commission their own versions of DOS.
Charles Simonyi and friends had been working up a suite of
MS-DOS applications with these varied platforms specifically in
mind. Multiplan, the spreadsheet, Multiword, later called just
Word, and all the other early Microsoft applications were
designed to be quickly ported to strange operating systems and
new hardware.

The idea was that Bill Gates would convince, say, Zenith, to
commission a custom version of MS-DOS. Once that project was
underway, it was time to remind Zenith that this new DOS version
might not work with all (or any) of the other DOS applications on
the market, most of which were customized for the IBM PC.

Panic time at Zenith headquarters in Illinois, where it became imperative to find some applications quickly that would
work with its new version of DOS. Son-of-a-gun, Microsoft *just
happened* to have a few portable applications lying around, written in a pseudocode that could be quickly adapted to almost any

165

computer. They weren't very good applications, but they were sure portable. And so Zenith, having been encouraged by Microsoft to do hardware incompatible with IBM's, then suckered into commissioning a custom version of MS-DOS, finally ended up having to pay Microsoft to adapt its applications, too. With all his costs covered, Bill Gates could start to make money even before the first copy of Multiplan or Word for Zenith was even sold.

This squeeze play happened for every new platform and every new version of MS-DOS and was just the first of many instances when Microsoft deliberately coordinated its operating system and application strategies, something the company continues to claim it never did.

As for the Victor 9000, the Z-100, the HP-150, the DEC Rainbow 100, and all the other early MS-DOS machines, those computers are gone now, dead and mainly forgotten. We can come up with all sorts of individual reasons why each machine failed, but at bottom they all failed because they were not IBM PC compatible. When the IBM PC, for all its faults, instantly became the number one selling personal computer, it became the de facto industry standard, because de facto standards are set by market share and nothing else. When Lotus 1-2-3 appeared, running on the IBM, and only on the IBM, the PC's role as the technical standard setter was guaranteed not just for this particular generation of hardware but for several generations of hardware.

The IBM PC defined what it meant to be a personal computer, and all these other computers that were sorta like the IBM PC, *kinda* like it, were doomed to eventual failure. They didn't even qualify as the requisite second standard—the pickup truck rather than the car—because although they were all different from the IBM PC, they weren't different enough to qualify for the number two spot.

Even the Grid Compass, the first laptop computer, was a failure because of a lack of IBM compatibility. Brilliant technology but different graphics and storage standards meant that Grid needed a version of 1-2-3 different from the one that worked on the IBM PC. When Grid supplied its own applications with the computer, including a spreadsheet, it still wasn't enough to attract buyers who wanted their 1-2-3. It was back to the drawing board to develop a second-generation laptop that was IBM compatible.

Entrepreneurs often lack the discipline to keep their new products tightly within a technical standard, which was why the idea of 100 percent IBM compatibility took so long to be accepted. "Why be compatible when you could be better?" the smart guys asked on their way to bankruptcy court.

IBM compatibility quickly became the key, and the level to which a computer was IBM compatible determined its success. Some long-established microcomputer makers learned this lesson slowly and expensively. Hewlett-Packard actually paid Lotus to adapt 1-2-3 to the HP-150, but the computer was still doomed by its lack of hardware compatibility (you couldn't put an IBM circuit card in an HP-150 computer). The other problem with the HP-150 was what was supposed to have been its major selling point—the touchscreen, which was a clever idea nobody really wanted. Not only was it hard to get software companies to make their products work with HP's touchscreen technology, users didn't like it. Secretaries, who apparently measure their self-worth by typing speed, didn't want to take their fingers off the keys. Even middle managers, who were the intended users of the system, didn't like the touch screen. The technology was clever, but it should have been a tip-off that H-P's own engineers chose not to use the systems. You could walk through the cavernlike open offices at H-P headquarters in those days without seeing a single user pointing at his or her touchscreen.

The best and most powerful computers come from designers who actually use their technologies—whose own tastes model those of intended users. Ivory towers, no matter how high, don't produce effective products for the real world.

Down at Tandy Corp. headquarters in Fort Worth, where ivory towers are unknown, Radio Shack's answer to the IBM PC was the Model 2000, another workalike, which appeared in the fall of 1983. The Model 2000 was intended to beat the IBM PC with twice the speed, more storage, and higher-resolution graphics. The trick was a more powerful processor, the Intel 80186, which could run rings around IBM's old 8088.

Because Tandy had its own distribution through 5,000 Radio Shack stores and through a chain of Tandy Computer Centers, the company thought for a long time that it was somehow immune to the influence of the IBM standard. They thought of their trusty Radio Shack customers as Albanians who would loyally shop at the Albanian Computer Store, no matter what was happening in the rest of the world. But Radio Shack's white-collar customer list turned out to include very few Albanians.

Bill Gates was a strong believer in the Model 2000 because it was the only personal computer powerful enough to run new software from Microsoft called Windows without being embarrassingly slow. Windows was an attempt to bring a Xerox Alto-style graphical user interface to personal computers. But Windows took a lot of power to run and was a real dog on the IBM PC and the other computers using 8088 processors. For Windows to succeed, Bill Gates needed a computer like the Model 2000. So young Bill, who handled the Tandy account himself, predicted that the computer would be a grand success—something the boys and girls in Fort Worth wanted badly to hear. And Gates made a public endorsement of the Model 2000, hoping to sway customers and promote Windows as well.

Still, the Model 2000 failed miserably. Nobody gave a damn about Windows, which didn't appear until 1985, and even then didn't work well. The computer wasn't hardware compatible with IBM. It wasn't very software compatible with IBM either, and the most popular IBM PC programs—the ones that talked directly to the PC's memory and so worked lots faster than those that allowed the operating system to do the talking for them—wouldn't work at all. Even the signals from the keyboard were different from IBM's, which drove software developers crazy and was one of the reasons that only a handful of software houses produced 2000-specific versions of their products. Oh, and the Intel 80186 processor had bugs, too, which took months to fix.

Today the Model 2000 is considered the magnum opus of Radio Shack marketing failures. Worse, a Radio Shack computer buyer in his last days with the company for some reason ordered 20,000 *more* of the systems built even when it was apparent they weren't selling. Tandy eventually sold over 5,000 of those systems to itself, placing one in each Radio Shack store to track inventory. Some leftover Model 2000s were still in the warehouse in early 1990, almost seven years later.

Still, the Model 2000's failure was Bill Gates's gain. Windows was a failure, but the head of Radio Shack's computer division, Jon Shirley, the very guy who'd been duped by Bill Gates into doing the Model 2000 in the first place, sensed that his position in Fort Worth was in danger and joined Microsoft as president in 1983.

Big Blue's share of the personal computer market peaked above 40 percent in the early 1980s. In 1983, IBM sold 538,000 personal computers. In 1984, it sold 1,375,000.

IBM wasn't afraid of others' copying the design of the PC, although nearly the entire system was built of off-the-shelf parts from other companies. Conventional wisdom in Boca Raton said that competitors would always pay more than IBM did for the parts needed to build a PC clone. To compete with IBM, another company would have to sell its PC clone at such a low price that there would be no profit. That was the theory.

In one sense, nothing could have been easier than building a PC clone, since IBM was so generous in supplying technical information about its systems. Everything a good engineer would need to know in order to design an IBM PC copy was readily available. While it seems like this would encourage copying, it was intended to do just the opposite because a trap lay in IBM's technical documentation. That trap was the complete code listing for the IBM PC's ROM-BIOS.

Remember, the ROM-BIOS was Gary Kildall's invention that allowed the same version of CP/M to operate on many different types of computers. The basic input/output system (BIOS) was special computer code that linked the generic operating system to specific hardware. The BIOS was stored in a read-only memory chip—a ROM—installed on the main computer circuit board, called the motherboard. To be completely compatible with the IBM PC, a clone machine either would have to use IBM's ROM-BIOS chip, which wasn't for sale, or devise another chip just like IBM's. But IBM's ROM-BIOS was copyrighted. The lines of code burned into the read-only memory were protected by law, so while it would be an easy matter to take IBM's published ROM-BIOS code and use it to prepare an exact copy of the chip, doing so would violate IBM's copyright and incur the legendary wrath of Armonk.

The key to making a copy of the IBM PC was copying the ROM-BIOS, and the key to copying the ROM-BIOS was to do so *without* reading IBM's published BIOS code.

Huh?

As we saw with Dan Bricklin's copyright on VisiCalc, a copyright protects only the specific lines of computer code but not the functions that those lines of code made the computer perform. The IBM copyright did not protect the company from others who might write their own completely independent code that just happened to perform the same BIOS function. By publishing its copyrighted BIOS code, IBM was making it very hard for others to claim that they had written their own BIOS without being exposed to or influenced by IBM's.

IBM was wrong. Welcome to the world of reverse engineering.

Reverse engineering is the science of copying a technical function without copying the legally protected manner in which that function is accomplished in a competitor's machine. Would-be PC clone makers had to come up with a chip that would replace IBM's ROM-BIOS but do so without copying any IBM code. The way this is done is by looking at IBM's ROM-BIOS as a black box —a mystery machine that does funny things to inputs and outputs. By knowing what data go into the black box—the ROM— and what data come out, programmers can make intelligent guesses about what happens to the data when they are inside the ROM. Reverse engineering is a matter of putting many of these guesses together and testing them until the cloned ROM-BIOS acts exactly like the target ROM-BIOS. It's a tedious and expensive process and one that can be accomplished only by virgins—programmers who can prove that they have never been exposed to IBM's ROM-BIOS code—and good virgins are hard to find.

Reverse engineering the IBM PC's ROM-BIOS took the efforts of fifteen senior programmers over several months and cost $1 million for the company that finally did it: Compaq Computer.

Compaq is the computer company with good penmanship. There was so little ego evident around the table when Rod Canion, Jim Harris, and Bill Murto were planning their start-up in the summer of 1981 that the three couldn't decide at first whether to open a Mexican restaurant, build hard disk drives for personal computers, or manufacture a gizmo that would beep on command to help find lost car keys. Oh, and they also considered starting a computer company. The computer company idea eventually won out, and the concept of the Compaq was first sketched out on a placemat at a House of Pies restaurant in Houston.

All three founders were experienced managers from Texas Instruments. TI was the company that many computer professionals expected throughout the late 1970s and early 1980s eventually to dominate the microcomputer business with its superior technology and management, only that never happened. Despite having the best chips, the brightest engineers, and Texas-sized ambition, the best TI did was a disastrous entry into the home computer business that eventually lost the company hundreds of millions of dollars. Later there was also an incompatible MS-DOS computer that came and went, suffering the same problem of attracting software as all the other rogue machines. Eventually TI produced a modest line of PC clones.

Unlike most of the other would-be IBM competitors, the three Compaq founders realized that software, and not hardware, was what really mattered. In order for their computer to be successful, it would have to have a large library of available software right from the start, which meant building a computer that was compatible with some other system. The only 16-bit standard available that qualified under these rules was IBM's, so that was the decision—to make an IBM-compatible PC—and to make it *damn* compatible—100 percent. Any program that would run on an IBM PC would run on a Compaq. Any circuit card that would operate in an IBM PC would operate in a Compaq. The key to their success would be leveraging the market's considerable investment in IBM.

Crunching the numbers harder than IBM had, the Compaq founders discovered that a smaller company with less overhead than IBM's could, in fact, bring out a lower-priced product and still make an acceptable profit. This didn't mean undercutting IBM by a lot but by a significant amount—about $800 on the first Compaq model compared to an IBM PC with equivalent features.

Compaq, like any other company pushing a new product, still had to ride the edges of an existing market, offering additional reasons for customers to choose its computer over IBM's. Just to be different, the first Compaq models were 28-pound portables— luggables, they came to be called. People didn't really drag these sewing machine–sized units around that much, but since IBM didn't make a luggable version of the PC, making theirs portable gave Compaq a niche to sell in right next to IBM.

Compaq appealed to computer dealers, even those who already sold IBM. *Especially* those who already sold IBM. For one thing, the Compaq portables were available, while IBM PCs were sometimes in short supply. Compaq pricing allowed dealers a 36 percent markup compared to IBM's 33 percent. And unlike IBM, Compaq had no direct sales force that competed with dealers. A third of IBM's personal computers were sold direct to major corporations, and each of those sales rankled some local dealer who felt cheated by Big Blue.

Just like IBM, Compaq first appeared in Sears Business Centers and ComputerLand stores, though a year later, at the end of 1982. With the Compaq's portability, compatibility, availability, and higher profit margins, signing up both chains was not difficult. Bill Murto made the ComputerLand sale by demonstrating the computer propped on the toilet seat in his hotel bathroom, the only place he could find a three-pronged electrical outlet.

Just like IBM, Compaq's dealer network was built by Sparky Sparks, who was hired away from Big Blue to do a repeat performance, selling similar systems to a virtually identical dealer network, though this time from Houston rather than Boca Raton.

By riding IBM's tail while being even better than IBM, Compaq sold 47,000 computers worth $111 million in its first year—a start-up record.

With the overnight success of Compaq, the idea of doing 100 percent IBM-compatible clones suddenly became very popular ("We'd intended to do it this way all along," the clone makers said), and the IBM workalikes quickly faded away. The most difficult and expensive part of Compaq's success had been developing the ROM-BIOS, a problem not faced by the many Compaq impersonators that suddenly appeared. What Compaq had done, companies like Phoenix Technologies could do too, and did. But Phoenix, a start-up from Boston, made its money not by building PC clones but by selling IBM-compatible BIOS chips to clone makers. Buying Phoenix's ROM-BIOS for $25 per chip, a couple of guys in a warehouse in New Jersey could put together systems that looked and ran just like IBM PCs, but cost 30 percent less to buy.

For months, IBM was shielded from the impact of the clone makers, first by Big Blue's own shortage of machines and later by a scam perpetrated by dealers.

When IBM's factories began churning out millions and millions of PCs, the computer giant set in place a plan that offered volume discounts to dealers. The more computers a dealer ordered, the less each computer cost. To make their cost of goods as low as possible, many dealers ordered as many computers as IBM would sell them, even if that was more computers than they could store at one time or even pay for. Having got the volume price, these dealers would sell excess computers out the back door to unauthorized dealers, at cost. Just when the planners in Boca Raton thought dealers were selling at retail everything they could make, these gray market PCs were being flogged by mail order or off the back of a truck in a parking lot, generally for 15 percent under list price.

Typical of these gray marketeers was Michael Dell, an 18-year-old student at the University of Texas with a taste for the finer things in life, who was soon clearing $30,000 per month selling gray market PCs from his Austin dorm room. Today Dell is a PC clone-maker, selling $400 million worth of IBM compatible computers a year.

Seeing this gray market scam as incessant demand, IBM just kept increasing production, increasing at the same time the downward pressure on gray market prices until some dealers were finally selling machines out of the back door for less than cost. That's when Big Blue finally noticed the clones.

For companies like IBM, the eventual problem with a hardware standard like the IBM PC is that it becomes a commodity. Companies you've never heard of in exotic places like Taiwan and Bayonne suddenly see that there is a big demand for specific PC power supplies, or cases, or floppy disk drives, or motherboards, and *whump!* the skies open and out fall millions of Acme power supplies, and Acme deluxe computer cases, and Acme floppy disk drives, and Acme Jr. motherboards, all built exactly like the ones used by IBM, just as good, and at one-third the price. It *always* happens. And if you, like IBM, are the caretaker of the hardware standard, or at least think that you still are, because sometimes such duties just drift away without their holder knowing it, the only way to fight back is by changing the rules. You've got to start selling a whole new PC that can't use Acme power supplies, or Acme floppy disk drives, or Acme Jr. motherboards, and just hope that the buyers will follow you to that new standard so the commoditization process can start all over again.

Commoditization is great for customers because it drives prices down and forces standard setters to innovate. In the absence of such competition, IBM would have done nothing. The company would still be building the original PC from 1981 if it could make enough profit doing so.

But IBM couldn't keep making a profit on its old hardware, which explains why Big Blue, in 1984, cut prices on its existing PC line and then introduced the PC-AT, a completely new computer that offered significantly higher performance and a certain amount of software compatibility with the old PC while conveniently having no parts in common with the earlier machine.

The AT was a speed demon. It ran two to three times faster than the old PCs and XTs. It had an Intel 80286 microprocessor, completely bypassing the flawed 80186 used in the Radio Shack Model 2000. Instead of a 360K floppy disk drive, the AT used a special 1.2-megabyte floppy, and every machine came with at least a 20-megabyte hard disk.

At around $4,000, the AT was also expensive, it wasn't able to run many popular PC-DOS applications, and sometimes it didn't run at all because the Computer Memories Inc. (CMI) hard disk used in early units had a tendency to die, taking the first ten chapters of your great American novel with it. IBM was so eager to swat Compaq and the lesser clone makers that it brought out the AT without adequate testing of the CMI drive's controller card built by Western Digital. There was no alternative controller to replace the faulty units, which led to months of angry customers and delayed production. Some customers who ordered the PC-AT at its introduction did not receive their machines for nine months.

The 80286 processor had been designed by Intel to operate in multi-user computers running a version of AT&T's Unix operating system called Xenix and sold by Microsoft. The chip was never intended to go in a PC. And in order to run Xenix efficiently, the 286 had two modes of operation—*real mode* and *protected mode*. In real mode, the 286 operated just like a very fast 8086 or 8088, and this was the way it could run some, but not all, MS-DOS applications. But protected mode was where the 286 showed its strength. In protected mode, the 286 could emulate several 8086s at once and could access vast amounts of memory. If real mode was impulse power, protected mode was warp speed. The only problem was that you couldn't get there from here.

The 286 chip powered up in real mode and then could be shifted into protected mode. This was the way Intel had envisoned it working in Xenix computers, which would operate strictly in protected mode. But the 286 was a chip that couldn't downshift; it could switch from real to protected mode but not from protected mode to real mode. The only way to get back to real mode was to turn the computer off, which was fine for a Xenix system at the end of the workday but pretty stupid for a PC that wanted to switch between a protected mode application and a real mode application. Until most applications ran in protected mode, then, the PC-AT would not reach its full potential.

And not only was the AT flawed, it was also late. The plan had been to introduce the new machine in early 1983, eighteen months after the original IBM PC and right in line with the trend of starting a new microcomputer generation every year and a half. But IBM's PC business unit was no longer able to bring a product to market in only eighteen months. They'd done the original PC in a year, but that had been in the time of gods, not men, before reality and the way that things have to be done in enormous companies had sunk in. Three years was how long it took IBM to invent a new computer, and the marketing staff in Boca Raton would just have to accept that and figure clever ways to keep the clones at bay for twice as long as they had been expected to before.

Still, the one-two punch of lowering PC prices and then introducing the AT took a toll on the clone makers, who had their already slim profit margins hurt by IBM's lower prices while simultaneously having to invest in cloning the AT.

The market loyally followed IBM to the AT standard, but life was never again as rosy for IBM as it had been in those earlier days of the original PC. Compaq, in a major effort, cloned the AT in only six months and shipped 10,000 of its Deskpro 286 models before IBM had solved the CMI drive problem and resumed its own AT shipments. But in the long term, Compaq was a small problem for IBM, compared to the one presented by Gordie Campbell.

Gordon Campbell was once the head of marketing at Intel. Like everyone else of importance at the monster chip company, he was an engineer. And as only an engineer could, one day Gordie fell in love with a new technology, the electrically erasable programmable read-only memory, or EEPROM, which doesn't mean beans to you or me but to computer engineers was a dramatic new type of memory chip that would make possible whole new categories of small-scale computer products. But where Gordie Campbell saw opportunity, the rest of Intel saw only a major technical headache because nobody had yet figured out how to manufacture EEPROMs in volume. Following a long Silicon Valley tradition, Campbell walked away from Intel, gathered up $30 million in venture capital, and started his EEPROM company—SEEQ Technologies. Who knows where they get these names?

With his $30 million, Campbell built SEEQ into a profitable company over the next four years, led the company through a successful public stock offering, and paid back the VCs their original investment, all without selling any EEPROMs, which were always three months away from being a viable technology. Still, SEEQ had its state-of-the-art chip-making facility and was able to make enough chips of other types to be profitable while continuing to tweak the EEPROM, which Campbell was sure would be ready Real Soon Now (a computer industry expression that means "in this lifetime, maybe").

Then one day Campbell came in to work at SEEQ's stylish headquarters only to find himself out of a job, fired by the company's lead venture capital firm, Kleiner Perkins Caulfield and Byers. Kleiner Perkins had the votes and Gordie, who held less than

3 percent of SEEQ stock, didn't, so he was out on the street, looking for his next start-up.

What happened to Campbell was that he came up against the fundamental conflict between venture capitalists and entrepreneurs. Like all other high-tech company founders, Campbell mistakenly assumed that Kleiner Perkins was investing in *his* dream, when, in fact, Kleiner Perkins was investing in *Kleiner Perkins's* dream, which just happened to involve Gordie Campbell. Sure SEEQ was already profitable and the VC's original investment had been repaid, but to an aggressive venture capitalist, that's just when real money starts to be made. And to Kleiner Perkins, it looked as if Gordie Campbell, for all his previous success, was making some bad decisions. Bye-bye, Gordie.

Campbell walked with $2 million in SEEQ stock, licked his wounds for a few months, and thought about his next venture. It had to be another chip company, he knew, but the question was whether to start a company to make general-purpose or custom semiconductors. General-purpose semiconductor companies like Intel, National Semiconductor, and Advanced Micro Devices took two to three years to develop chips, which were then sold in the millions for use in all sorts of electronic equipment. Custom chip companies developed their products in only a few months through the use of expensive computer design tools, with the result being high-performance chips that were sold in very small volumes, mainly to defense contractors at astronomical prices.

Campbell decided to follow an edge of the market. He would apply to general-purpose chip development the computer-intensive design tools of the custom semiconductor makers. Just as Compaq could produce a new computer in six months, Campbell wanted to start a semiconductor company that could develop new chips in that amount of time and then sell millions of them to the personal computer industry.

The investment world was doubtful. Becoming increasingly

convinced that he had been blackballed by Kleiner Perkins, Campbell traveled the world looking for venture capital. His pitch was rejected sixty times. The new company, Chips & Technologies, finally got moving on $1.5 million from Campbell and a friend who was a real estate developer. Nearly all the money went into leasing giant IBM 3090 and Amdahl 470 mainframes used to design the new chips. When that money was gone, Campbell depleted his savings and then borrowed from his chief financial officer to make payroll. Broke again, and with still no chip designs completed, he finally went to the Far East to look for money, financing the trip on his credit cards. On his last day abroad, Campbell met with Kay Nishi, who then represented Microsoft in Japan. Nishi put together a group of Japanese investors who came up with another $1.5 million in exchange for 15 percent of the company. This was all the money Chips & Technologies ever raised—$3 million total.

At SEEQ, most of the $30 million in venture capital had been spent building a semiconductor factory. That's the way it was with chip companies, where everyone thought that they could do a better job than the other guys at making chips. But Chips & Technologies couldn't afford to build a factory. Then Campbell discovered that all the chip makers with edifice complexes had produced a glut of semiconductor production capacity. He could farm out his chip production cheaper than doing it in-house.

As always, the real value lay in the design—in software—not in hardware. There was nothing sacred about a factory.

The first C&T product was a set of five chips that hit the market in the fall of 1985. These five chips, which sold then for $72.40, replaced sixty-three smaller chips on an IBM PC-AT motherboard. Using the C&T chip set, clone makers could build a 100 percent IBM-compatible AT clone with 256K of memory using only twenty-four chips. They could *buy* 100 percent IBM compatibility. Their personal computers could suddenly be smaller, easier to build, more reliable, even faster than a real IBM AT. And because they weren't having to buy all the same individual parts as IBM, the clone makers could put together AT clones for less than it cost IBM, even with Big Blue's massive buying power, to build the real thing.

Chips & Technologies was an overnight success, getting the world back on the traditional track of computers' doubling in power and halving in price every eighteen months. Venture capital firms—the same ones that rejected Campbell sixty times in a row—immediately funded half a dozen companies just like Chips.

The commoditization of the PC-AT was complete, and though it didn't know it at the time, IBM had lost forever its control of the personal computer business.

THE PROPHET

The most dangerous man in Silicon Valley sits alone on many weekday mornings, drinking coffee at Il Fornaio, an Italian restaurant on Cowper Street in Palo Alto. He's not the richest guy around or the smartest, but under a haircut that looks as if someone put a bowl on his head and trimmed around the edges, Steve Jobs holds an idea that keeps some grown men and women of the Valley awake at night. Unlike these insomniacs, Jobs isn't in this business for the money, and *that's* what makes him dangerous.

I wish, sometimes, that I could say this personal computer stuff is just a matter of hard-headed business, but that would in no way account for the phenomenon of Steve Jobs. Co-founder of Apple Computer and founder of NeXT Inc., Jobs has literally forced the personal computer industry to follow his direction for fifteen years, a direction based not on business or intellectual principles but on a combination of technical vision and ego gratification in which both business and technical acumen played only small parts.

Steve Jobs sees the personal computer as his tool for changing the world. I know that sounds a lot like Bill Gates, but it's really very different. Gates sees the personal computer as a tool for transferring every stray dollar, deutsche mark, and kopeck in the world into his pocket. Gates doesn't really give a damn how people interact with their computers as long as they pay up. Jobs gives a damn. He wants to tell the world how to compute, to set the style for computing.

Bill Gates has no style; Steve Jobs has nothing but style.

A friend once suggested that Gates switch to Armani suits from his regular plaid shirt and Levis Dockers look. "I can't do that," Bill replied. "Steve Jobs wears Armani suits."

Think of Bill Gates as the emir of Kuwait and Steve Jobs as Saddam Hussein.

Like the emir, Gates wants to run his particular subculture with an iron hand, dispensing flawed justice as he sees fit and generally keeping the bucks flowing in, not out. Jobs wants to control the world. He doesn't care about maintaining a strategic advantage; he wants to attack, to bring death to the infidels. We're talking rivers of blood here. We're talking martyrs. Jobs doesn't care if there are a dozen companies or a hundred companies opposing him. He doesn't care what the odds are against success. Like Saddam, he doesn't even care how much his losses are. Nor does he even *have* to win, if, by losing the mother of all battles he can maintain his peculiar form of conviction, still stand before an adoring crowd of nerds, symbolically firing his 9 mm automatic into the air, telling the victors that they are still full of shit.

You guessed it. By the usual standards of Silicon Valley CEOs, where job satisfaction is measured in dollars, and an opulent retirement by age 40 is the goal, Steve Jobs is crazy.

Apple Computer was always different. The company tried hard from the beginning to shake the hobbyist image, replacing it with the idea that the Apple II was an appliance but not just any appliance; it was the next great appliance, a Cuisinart for the mind. Apple had the five-color logo and the first celebrity spokesperson: Dick Cavett, the thinking person's talk show host.

Alone among the microcomputer makers of the 1970s, the people of Apple saw themselves as not just making boxes or making money; they thought of themselves as changing the world.

Atari wasn't changing the world; it was in the entertainment business. Commodore wasn't changing the world; it was just trying to escape from the falling profit margins of the calculator market while running a stock scam along the way. Radio Shack wasn't changing the world; it was just trying to find a new consumer wave to ride, following the end of the CB radio boom. Even IBM, which already controlled the world, had no aspirations to change it, just to wrest some extra money from a small part of the world that it had previously ignored.

In contrast to the hardscrabble start-ups that were trying to eke out a living selling to hobbyists and experimenters, Apple was appealing to doctors, lawyers, and middle managers in large corporations by advertising on radio and in full pages of *Scientific American.* Apple took a heroic approach to selling the personal computer and, by doing so, taught all the others how it should be done.

They were heroes, those Apple folk, and saw themselves that way. They were more than a computer company. In fact, to figure out what was going on in the upper echelons in those Apple II days, think of it not as a computer company at all but as an episode of "Bonanza."

(*Theme music, please.*)

Riding straight off the Ponderosa's high country range every Sunday night at nine was Ben Cartwright, the wise and supportive father, who was willing to wield his immense power if needed. At Apple, the part of Ben was played by Mike Markkula.

Adam Cartwright, the eldest and best-educated son, who was sophisticated, cynical, and bossy, was played by Mike Scott. Hoss Cartwright, a good-natured guy who was capable of amazing feats of strength but only when pushed along by the others, was played by Steve Wozniak. Finally, Little Joe Cartwright, the baby of the family who was quick with his mouth, quick with his gun, but was never taken as seriously as he wanted to be by the rest of the family, was played by young Steve Jobs.

The series was stacked against Little Joe. Adam would always be older and more experienced. Hoss would always be stronger. Ben would always have the final word. Coming from this environment, it was hard for a Little Joe character to grow in his own right, short of waiting for the others to die. Steve Jobs didn't like to wait.

By the late 1970s, Apple was scattered across a dozen one- and two-story buildings just off the freeway in Cupertino, California. The company had grown to the point where, for the first time, employees didn't all know each other on sight. Maybe that kid in the KOME T-shirt who was poring over the main circuit board of Apple's next computer was a new engineer, a manufacturing guy, a marketer, or maybe he wasn't any of those things and had just wandered in for a look around. It had happened before. Worse, maybe he was a spy for the other guys, which at that time didn't mean IBM or Compaq but more likely meant the start-up down the street that was furiously working on its own micro-computer, which its designers were sure would soon make the world forget that there ever was a company called Apple.

Facing these realities of growth and competition, the grown-ups at Apple—Mike Markkula, chairman, and Mike Scott, presi-dent—decided that ID badges were in order. The badges included a name and an individual employee number, the latter based on the order in which workers joined the company. Steve Wozniak was declared employee number 1, Steve Jobs was number 2, and so on.

Jobs didn't want to be employee number 2. He didn't want to be second in anything. Jobs argued that he, rather than Woz, should have the sacred number 1 since they were co-founders of the company and J came before W in the alphabet. It was a kid's argument, but then Jobs, who was still in his early twenties, *was* a kid. When that plan was rejected, he argued that the number 0 was still unassigned, and since 0 came before 1, Jobs would be happy to take that number. He got it.

Steve Wozniak deserved to be considered Apple's number 1 employee. From a technical standpoint, Woz literally *was* Apple Computer. He designed the Apple II and wrote most of its system software and its first BASIC interpreter. With the exception of the computer's switching power supply and molded plastic case, literally every other major component in the Apple II was a product of Wozniak's mind and hand.

And in many ways, Woz was even Apple's conscience. When the company was up and running and it became evident that some early employees had been treated more fairly than others in the distribution of stock, it was Wozniak who played the peacemaker, selling cheaply 80,000 of his own Apple shares to employees who felt cheated and even to those who just wanted to make money at Woz's expense.

Steve Jobs's roles in the development of the Apple II were those of purchasing agent, technical gadfly, and supersalesman. He nagged Woz into a brilliant design performance and then took Woz's box to the world, where through sheer force of will, this kid with long hair and a scraggly beard imprinted his enthusiasm for the Apple II on thousands of would-be users met at computer shows. But for all Jobs did to sell the world on the idea of buying a microcomputer, the Apple II would always be Wozniak's machine, a fact that might have galled employee number 0, had he allowed it to. But with the huckster's eternal optimism, Jobs was always looking ahead to the next technical advance, the next computer, determined that that machine would be all his.

Jobs finally got the chance to overtake his friend when Woz was hurt in the February 1981 crash of his Beechcraft Bonanza after an engine failure taking off from the Scotts Valley airport. With facial injuries and a case of temporary amnesia, Woz was away from Apple for more than two years, during which he returned to Berkeley to finish his undergraduate degree and produced two rock festivals that lost a combined total of nearly $25 million, proving that not everything Steve Wozniak touched turned to gold.

Another break for Jobs came two months after Woz's airplane crash, when Mike Scott was forced out as Apple president, a victim of his own ruthless drive that had built Apple into a $300 million company. Scott was dogmatic. He did stupid things like issuing edicts against holding conversations in aisles or while standing. Scott was brusque and demanding with employees ("Are you working your ass off?" he'd ask, glaring over an office cubicle partition). And when Apple had its first-ever round of layoffs, Scott handled them brutally, pushing so hard to keep momentum going that he denied the company a chance to mourn its loss of innocence.

Scott was a kind of clumsy parent who tried hard, sometimes too hard, and often did the wrong things for the right reasons. He was not well suited to lead the $1 billion company that Apple would soon be.

Scott had carefully thwarted the ambitions of Steve Jobs. Although Jobs owned 10 percent of Apple, outside of purchasing (where Scott still insisted on signing the purchase orders, even if Jobs negotiated the terms), he had little authority.

Mike Markkula fired Scott, sending the ex-president into a months-long depression. And it was Markkula who took over as president when Scott left, while Jobs slid into Markkula's old job as chairman. Markkula, who'd already retired once before, from Intel, didn't really want the president's job and in fact had been trying to remove himself from day-to-day management responsibility at Apple. As a president with retirement plans, Markkula was easier-going than Scott had been and looked much more kindly on Jobs, whom he viewed as a son.

Every high-tech company needs a technical visionary, someone who has a clear idea about the future and is willing to do whatever it takes to push the rest of the operation in that direction. In the earliest days of Apple, Woz was the technical visionary along with doing nearly everything else. His job was to see the potential product that could be built from a pile of computer chips. But that was back when the world was simpler and the paradigm was to bring to the desktop something that emulated a mainframe computer terminal. After 1981, Woz was gone, and it was time for someone else to take the visionary role. The only people inside Apple who really wanted that role were Jef Raskin and Steve Jobs.

Raskin was an iconoclastic engineer who first came to Apple to produce user manuals for the Apple II. His vision of the future was a very basic computer that would sell for around $600—a computer so easy to use that it would require no written instructions, no user training, and no product support from Apple. The new machine would be as easy and intuitive to use as a toaster and would be sold at places like Sears and K-Mart. Raskin called his computer Macintosh.

Jobs's ambition was much grander. He wanted to lead the development of a radical and complex new computer system that featured a graphical user interface and mouse (Raskin preferred keyboards). Jobs's vision was code-named Lisa.

Depending on who was talking and who was listening, Lisa was either an acronym for "large integrated software architecture," or for "local integrated software architecture" or the name of a daughter born to Steve Jobs and Nancy Rogers in May 1978. Jobs, the self-centered adoptee who couldn't stand competition from a baby, at first denied that he was Lisa's father, sending mother and baby for a time onto the Santa Clara County welfare rolls. But blood tests and years later, Jobs and Lisa, now a teenager, are often seen rollerblading on the streets of Palo Alto. Jobs and Rogers never married.

Lisa, the computer, was born after Jobs toured Xerox PARC in December 1979, seeing for the first time what Bob Taylor's crew at the Computer Science Lab had been able to do with bitmapped video displays, graphical user interfaces, and mice. "Why aren't you marketing this stuff?" Jobs asked in wonderment as the Alto and other systems were put through their paces for him by a PARC scientist named Larry Tesler. Good question.

Steve Jobs saw the future that day at PARC and decided that if Xerox wouldn't make that future happen, then he would. Within days, Jobs presented to Markkula his vision of Lisa, which included a 16-bit microprocessor, a bit-mapped display, a mouse for controlling the on-screen cursor, and a keyboard that was separate from the main computer box. In other words, it was a Xerox Alto, minus the Alto's built-in networking. "Why would anyone need an umbilical cord to his company?" Jobs asked.

Lisa was a vision that made the as-yet-unconceived IBM PC look primitive in comparison. And though he didn't know it at the time, it was also a development job far bigger than Steve Jobs could even imagine.

One of the many things that Steve Jobs didn't know in those days was Cringely's Second Law, which I figured out one afternoon with the assistance of a calculator and a six-pack of Heineken. Cringely's Second Law states that in computers, ease of use with equivalent performance varies with the square root of the cost of development. This means that to design a computer that's ten times easier to use than the Apple II, as the Lisa was intended to be, would cost 100 times as much money. Since it cost around $500,000 to develop the Apple II, Cringely's Second Law says the cost of building the Lisa should have been around $50 million. It was.

Let's pause the history for a moment and consider the implications of this law for the *next* generation of computers. There was no significant difference in ease of use between Lisa and its follow-on, the Macintosh. So if you've been sitting on your hands waiting to buy a computer that is ten times as easy to use as the Macintosh, remember that it's going to cost around $5 billion (1982 dollars, too) to develop. Apple's R&D budget is about $500 million, so don't expect that computer to come from Cupertino. IBM's R&D budget is about $3 billion, but that's spread across many lines of computers, so don't expect your ideal machine to come from Big Blue either. The only place such a computer is going to come from, in fact, is a collaboration of computer and semiconductor companies. That's why the computer world is suddenly talking about *Open Systems,* because building hardware and software that plug and play across the product lines and R&D budgets of a hundred companies is the only way that future is going to be born. Such collaboration, starting now, will be the trend in the next century, so put your wallet away for now.

Meanwhile, back in Cupertino, Mike Markkula knew from his days working in finance at Intel just how expensive a big project could become. That's why he chose John Couch, a software professional with a track record at Hewlett-Packard, to head the super-secret Lisa project. Jobs was crushed by losing the chance to head the realization of his own dream.

Couch was yet another Adam Cartwright, and Jobs hated him.

The new ideas embodied in Lisa would have been Jobs's way

of breaking free from his type casting as Little Joe. He would become, instead, the prophet of a new kind of computing, taking his power from the ideas themselves and selling this new type of computing to Apple and to the rest of the world. And Apple accepted both his dream and the radical philosophy behind it, which said that technical leadership was as important as making money, but Markkula still wouldn't let him lead the project.

Vision, you'll recall, is the ability to see potential in the work of others. The jump from having vision to being a visionary, though, is a big one. The visionary is a person who has both the vision and the willingness to put everything on the line, including his or her career, to further that vision. There aren't many real visionaries in this business, but Steve Jobs is one. Jobs became the perfect visionary, buying so deeply into the vision that he became one with it. If you were loyal to Steve, you embraced his vision. If you did not embrace his vision, you were either an enemy or brain-dead.

So Chairman Jobs assigned himself to Raskin's Macintosh group, pushed the other man aside, and converted the Mac into what was really a smaller, cheaper Lisa. As the holder of the original Lisa vision, Jobs ignorantly criticized the big-buck approach being taken by Couch and Larry Tesler, who had by then joined Apple from Xerox PARC to head Lisa software development. Lisa was going to be too big, too slow, too expensive, Jobs argued. He bet Couch $5,000 that Macintosh would hit the market first. He lost.

The early engineers were nearly all gone from Apple by the time Lisa development began. The days when the company ran strictly on adrenalin and good ideas were fading. No longer did the whole company meet to put computers in boxes so they could ship enough units by the end of the month. With the introduction of the Apple III in 1980, life had become much more businesslike at Apple, which suddenly had two product lines to sell.

It was still the norm, though, for technical people to lead each product development effort, building products that they wanted to play with themselves rather than products that customers wanted to buy. For example, there was Mystery House, Apple's own spreadsheet, intended to kill VisiCalc because everyone who worked on Apple II software decided en masse that they hated Terry Opdendyk, president of VisiCorp, and wanted to hurt him by destroying his most important product. There was no real business reason to do Mystery House, just spite. The spreadsheet was written by Woz and Randy Wigginton and never saw action under the Apple label because it was given up later as a bargaining chip in negotiations between Apple and Microsoft. Some Mystery House code lives on today in a Macintosh spreadsheet from Ashton-Tate called Full Impact.

But John Couch and his Lisa team were harbingers of a new professionalism at Apple. Apple had in Lisa a combination of the old spirit of Apple—anarchy, change, new stuff, engineers working through the night coming up with great ideas—and the introduction of the first nontechnical marketers, marketers with business degrees—the "suits." These nontechnical marketers were, for the first time at Apple, the project coordinators, while the technical people were just members of the team. And rather than the traditional bunch of hackers from Homestead High, Lisa hardware was developed by a core of engineers hired away from Hewlett-Packard and DEC, while the software was developed mainly by ex-Xerox programmers, who were finally getting a chance to bring to market a version of what they'd worked on at Xerox PARC for most of the preceding ten years. Lisa was the most professional operation ever mounted at Apple—far more professional than anything that has followed.

Lisa was ahead of its time. When most microcomputers

came with a maximum of 64,000 characters of memory, the Lisa had 1 million characters. When most personal computers were capable of doing only one task at a time, Lisa could do several. The computer was so easy to use that customers were able to begin working within thirty minutes of opening the box. Setting up the system was so simple that early drafts of the directions used only pictures, no words. With its mouse, graphical user interface, and bit-mapped screen, Lisa was the realization of nearly every design feature invented at Xerox PARC except networking.

Lisa was professional all the way. Painstaking research went into every detail of the user interface, with arguments ranging up and down the division about what icons should look like, whether on-screen windows should just appear and disappear or whether they should zoom in and out. Unlike nearly every other computer in the world, Lisa had no special function keys to perform complex commands in a single keystroke, and offered no obscure ways to hold down three keys simultaneously and, by so doing, turn the whole document into Cyrillic, or check its spelling, or some other such nonsense.

To make it easy to use, Lisa followed PARC philosophy, which meant that no matter what program you were using, hitting the E key just put an E on-screen rather than sending the program into edit mode, or expert mode, or erase mode. Modes were evil. At PARC, you were either modeless or impure, and this attitude carried over to Lisa, where Larry Tesler's license plate read NO MODES. Instead of modes, Lisa had a very simple keyboard that was used in conjunction with the mouse and on-screen menus to manipulate text and graphics without arcane commands.

Couch left nothing to chance. Even the problem of finding a compelling application for Lisa was covered; instead of waiting for a Dan Bricklin or a Mitch Kapor to introduce the application that would make corporate America line up to buy Lisas, Apple wrote its own software—seven applications covering everything that users of microcomputers were then doing with their machines, including a powerful spreadsheet.

Still, when Lisa hit the market in 1983, it failed. The problem was its $10,000 price, which meant that Lisa wasn't really a personal computer at all but the first real workstation. Workstations can cost more than PCs because they are sold to companies rather than to individuals, but they have to be designed with companies in mind, and Lisa wasn't. Apple had left out that umbilical cord to the company that Steve Jobs had thought unnecessary. At $10,000, Lisa was being sold into the world of corporate mainframes, and the Apple's inability to communicate with those mainframes doomed it to failure.

Despite the fact that Lisa had been his own dream and Apple was his company, Steve Jobs was thrilled with Lisa's failure, since it would make the inevitable success of Macintosh all the more impressive.

Back in the Apple II and Apple III divisions, life still ran at a frenetic pace. Individual contributors made major decisions and worked on major programs alone or with a very few other people. There was little, if any, management, and Apple spent so much money, it was unbelievable. With Raskin out of the way, that's how Steve Jobs ran the Macintosh group too. The Macintosh was developed beneath a pirate flag. The lobby of the Macintosh building was lined with Ansel Adams prints, and Steve Jobs's BMW motorcycle was parked in a corner, an ever-present reminder of who was boss. It was a renegade operation and proud of it.

When Lisa was taken from him, Jobs went through a paradigm shift that combined his dreams for the Lisa with Raskin's idea of appliancelike simplicity and low cost. Jobs decided that the problem with Lisa was not that it lacked networking capability but that its high price doomed it to selling in a market that demanded networking. There'd be no such problem with Macintosh, which would do all that Lisa did but at a vastly lower price. Never mind that it was technically impossible.

Lisa was a big project, while Macintosh was much smaller because Jobs insisted on an organization small enough that he could dominate every member, bending each to his will. He built the Macintosh on the backs of Andy Hertzfeld, who wrote the system software, and Burrell Smith, who designed the hardware. All three men left their idiosyncratic fingerprints all over the machine. Hertzfeld gave the Macintosh an elegant user interface and terrific look and feel, mainly copied from Lisa. He also made Macintosh very, very difficult to write programs for. Smith was Jobs's ideal engineer because he'd come up from the Apple II service department ("I made him," Jobs would say). Smith built a clever little box that was incredibly sophisticated and nearly impossible to manufacture.

Jobs's vision imposed so many restraints on the Macintosh that it's a wonder it worked at all. In contrast to Lisa, with its million characters of memory, Raskin wanted Macintosh to have only 64,000 characters—a target that Jobs continued to aim for until long past the time when it became clear to everyone else that the machine needed more memory. Eventually, he "allowed" the machine to grow to 128,000 characters, though even with that amount of memory, the original 128K Macintosh still came to fit people's expectations that mechanical things don't work. Apple engineers, knowing that further memory expansion was inevitable, built in the capability to expand the 128K machine to 512K, though they couldn't tell Jobs what they had done because he would have made them change it back.

Markkula gave up the presidency of Apple at about the time Lisa was introduced. As chairman, Jobs went looking for a new president, and his first choice was Don Estridge of IBM, who turned the job down. Jobs's second choice was John Sculley, who came over from PepsiCo for the same package that Estridge had rejected. Sculley was going to be as much Jobs's creation as Burrell Smith had been. It was clear to the Apple technical staff that Sculley knew nothing at all about computers or the computer business. They dismissed him, and nobody even noticed when Sculley was practically invisible during his first months at Apple. They thought of him as Jobs's lapdog, and that's what he was.

With Mike Markkula again in semiretirement, concentrating on his family and his jet charter business, there was no adult supervision in place at Apple, and Jobs ran amok. With total power, the willful kid who'd always resented the fact that he had been adopted, created at Apple a metafamily in which he played the domineering, disrespectful, demanding type of father that he imagined must have abandoned him those many years ago.

Here's how Steve-As-Dad interpreted Management By Walking Around. Coming up to an Apple employee, he'd say, "I think Jim [another employee] is shit. What do you think?"

If the employee agrees that Jim is shit, Jobs went to the next person and said, "Bob and I think Jim is shit. What do you think?"

If the first employee disagreed and said that Jim is not shit, Jobs would move on to the next person, saying, "Bob and I think Jim is great. What do you think?"

Public degradation played an important role too. When Jobs finally succeeded in destroying the Lisa division, he spoke to the assembled workers who were about to be reassigned or laid off. "I see only B and C players here," he told the stunned assemblage. "All the A players work for me in the Macintosh division. I might be interested in hiring two or three of you [out of 300]. Don't you wish you knew which ones I'll choose?"

Jobs was so full of himself that he began to believe his own PR, repeating as gospel stories about him that had been invented to help sell computers. At one point a marketer named Dan'l Lewin stood up to him, saying, "Steve, we wrote this stuff about you. We *made it up*."

Somehow, for all the abuse he handed out, nobody attacked Jobs in the corridor with a fire axe. I would have. Hardly anyone stood up to him. Hardly anyone quit. Like the Bhagwan, driving around Rancho Rajneesh each day in another Rolls-Royce, Jobs kept his troops fascinated and productive. The joke going around said that Jobs had a "reality distortion field" surrounding him. He'd say something, and the kids in the Macintosh division would find themselves replying, "Drink poison Kool-Aid? Yeah, that makes sense."

Steve Jobs gave impossible tasks, never acknowledging that they were impossible. And, as often happens with totalitarian rulers, most of his impossible demands were somehow accomplished, though at a terrible cost in ruined careers and failed marriages.

Beyond pure narcissism, which was there in abundance, Jobs used these techniques to make sure he was surrounding himself with absolutely the best technical people. The best, nothing but the best, was all he would tolerate, which meant that there were crowds of less-than-godlike people who went continually up and down in Jobs's estimation, depending on how much he needed them at that particular moment. It was crazy-making.

Here's a secret to getting along with Steve Jobs: when he screams at you, scream back. Take no guff from him, and if he's the one who is full of shit, tell him, preferably in front of a large group of amazed underlings. This technique works because it gets Jobs's attention and fits in with his underlying belief that he probably *is* wrong but that the world just hasn't figured that out yet. Make it clear to him that you, at least, know the truth.

Jobs had all kinds of ideas he kept throwing out. Projects would stop. Projects would start. Projects would get so far and then be abandoned. Projects would go on in secret, because the budget was so large that engineers could hide things they wanted to do, even though that project had been canceled or never approved. For example, Jobs thought at one point that he had killed the Apple III, but it went on anyhow.

Steve Jobs created chaos because he would get an idea, start a project, then change his mind two or three times, until people were doing a kind of random walk, continually scrapping and starting over. Apple was confusing suppliers and wasting huge amounts of money doing initial manufacturing steps on products that never appeared.

Despite the fact that Macintosh was developed with a much smaller team than Lisa and it took advantage of Lisa technology, the little computer that was supposed to have sold at K-Mart for $600 ended up costing just as much to bring to market as Lisa had. From $600, the price needed to make a MacProfit doubled and tripled until the Macintosh could no longer be imagined as a home computer. Two months before its introduction, Jobs declared the Mac to be a business computer, which justified the higher price.

Apple clearly wasn't very disciplined. Jobs created some of that, and a lot of it was created by the fact that it didn't matter to him whether things were organized. Apple people were rewarded for having great ideas and for making great technical contributions but not for saving money. Policies that looked as if they were aimed at saving money actually had other justifications. Apple people still share hotel rooms at trade shows and company meetings, for example, but that's strictly intended to limit bed hopping, not to save money. Apple is a very sexy company, and Jobs wanted his people to lavish that libido on the products rather than on each other.

Oh, and Apple people were also rewarded for great graphics; brochures, ads, everything that represented Apple to its customers and dealers, had to be absolutely top quality. In addition, the people who developed Apple's system of dealers were rewarded because the company realized early on that this was its major strength against IBM.

A very dangerous thing happened with the introduction of the Macintosh. Jobs drove his development team into the ground, so when the Mac was introduced in 1984, there was no energy left, and the team coasted for six months and then fell apart. And during those six months, John Sculley was being told that there were development projects going on in the Macintosh group that weren't happening. The Macintosh people were just burned out, the Lisa Division was destroyed and its people were not fully integrated into the Macintosh group, so there was no new blood.

It was a time when technical people should have been fixing the many problems that come with the first version of any complex high-tech product. But nobody moved quickly to fix the problems. They were just too tired.

The people who made the Macintosh produced a miracle, but that didn't mean their code was wonderful. The software development tools to build applications like spreadsheets and word processors were not available for at least two years. Early Macintosh programs had to be written first on a Lisa and then recompiled to run on the Mac. None of this mattered to Jobs, who was in heaven, running Apple as his own private psychology experiment, using up people and throwing them away. Attrition, strangled marriages, and destroyed careers were unimportant, given the broader context of his vision.

The idea was to have a large company that somehow maintained a start-up philosophy, and Jobs thrived on it. He planned to develop a new generation of products every eighteen months, each one as radically different from the one before as the Macintosh had been from the Apple II. By 1990, nobody would even remember the Macintosh, with Apple four generations down the road. Nothing was sacred except the vision, and it became clear to him that the vision could best be served by having the people of Apple live and work in the same place. Jobs had Apple buy hundreds of acres in the Coyote Valley, south of San Jose, where he planned to be both employer and landlord for his workers, so they'd never ever have a reason to leave work.

Unchecked, Jobs was throwing hundreds of millions of dollars at his dream, and eventually the drain became so bad that Mike Markkula revived his Ben Cartwright role in June 1985. By this point Sculley had learned a thing or two in his lapdog role and felt ready to challenge Jobs. Again, Markkula decided against Jobs, this time backing Sculley in a boardroom battle that led to Jobs's being banished to what he called "Siberia"— Bandley 6, an Apple building with only one office. It was an office for Steve Jobs, who no longer had any official duties at the company he had founded in his parents' garage. Jobs left the company soon after.

Here's what was happening at Apple in the early 1980s that Wall Street analysts didn't know. For its first five years in business, Apple did not have a budget. Nobody really knew how much money was coming in or going out or what the company was buying. In the earliest days, this wasn't a problem because a company that was being run by characters who not long before had made $3 per hour dressing up as figures from *Alice in Wonderland* at a local shopping mall just wasn't inclined toward extravagance. Later, it seemed that the money was coming in so fast that there was no way it could all be spent. In fact, when the first company budget happened in 1982, the explanation was that Apple finally had enough people and projects where they could actually spend all the money they made if they didn't watch it.

But even when they got a budget, Apple's budgeting process

was still a joke. All budgets were done at the same time, so rather than having product plans from which support plans and service plans would flow—a logical plan based on products that were coming out—everybody all at once just said what they wanted. Nothing was coordinated.

It really wasn't until 1985 that there was any logical way of making the budget, where the product people would say what products would come out that year, and then the marketing people would say what they were going to do to market these products, and the support people would say how much it was going to cost to support the products.

It took Sculley at least six months, maybe a year, from the time he deposed Jobs to understand how out of control things were. It was total anarchy. Sculley's major budget gains in the second half of 1985 came from laying off 20 percent of the work force—1,200 people—and forcing managers to make sense of the number of suppliers they had and the spare parts they had on hand. Apple had millions of dollars of spare parts that were never going to be used, and many of these were sold as surplus. Sculley instituted some very minor changes in 1986—reducing the number of suppliers and beginning to simplify the peripherals line so that Macintosh printers, for example, would also work with the Apple II, Apple III, and Lisa.

The large profits that Sculley was able to generate during this period came entirely from improved budgeting and from simply cancelling all the whacko projects started by Steve Jobs. Sculley was no miracle worker.

Who was this guy Sculley? Raised in Bermuda, scion of an old-line, old-money family, he trained as an architect, then worked in marketing at PepsiCo for his entire career before joining Apple. A loner, his specialty at the soft drink maker seemed to be corporate infighting, a habit he brought with him to Apple.

Sculley is not an easy man to be with. He is uneasy in public and doesn't fit well with the casual hacker class that typified the Apple of Woz and Jobs. Spend any time with Sculley and you'll notice his eyes, which are dark, deep-set, and hawklike, with white visible on both sides of the iris and above it when you look at him straight on. In traditional Japanese medicine, where facial features are used for diagnosis, Sculley's eyes are called *sanpaku* and are attributed to an excess of yang. It's a condition that Japanese doctors associate with people who are prone to violence.

With Jobs gone, Apple needed a new technical visionary. Sculley tried out for the role, and supported people like Bill Atkinson, Larry Tesler, and Jean-Louis Gassée as visionaries, too. He tried to send a message to the troops that everything would be okay, and that wonderful new products would continue to come out, except in many ways they didn't.

Sculley and the others were surrogate visionaries compared to Jobs. Sculley's particular surrogate vision was called Knowledge Navigator, mapped out in an expensive video and in his book, *Odyssey*. It was a goal, but not a product, deliberately set in the far future. Jobs would have set out a vision that he intended his group actually to accomplish. Sculley didn't do that because he had no real goal.

By rejecting Steve Jobs's concept of continuous revolution but not offering a specific alternative program in its place, Sculley was left with only the status quo. He saw his job as milking as much money as possible out of the current Macintosh technology and allowing the future to take care of itself. He couldn't envision later generations of products, and so there would be none. Today the Macintosh is a much more powerful machine, but it still has an operating system that does only one thing at a time. It's the same old stuff, only faster.

And along the way, Apple abandoned the $1-billion-per-year

Apple II business. Steve Jobs had wanted the Apple II to die because it wasn't his vision. Then Jean-Louis Gassée came in from Apple France and used his background in minicomputers to claim that there really wasn't a home market for personal computers. Earth to Jean-Louis! Earth to Jean-Louis! So Apple ignored the Macintosh home market to develop the Macintosh business market, and all the while, the company's market share continued to drop.

Sculley didn't have a clue about which way to go. And like Markkula, he faded in and out of the business, residing in his distant tower for months at a time while the latest group of subordinates would take their shot at running the company. Sculley is a smart guy but an incredibly bad judge of people, and this failing came to permeate Apple under his leadership.

Sculley falls in love with people and gives them more power than they can handle. He chose Gassée to run Apple USA and the phony-baloney Frenchman caused terrific damage during his tenure. Gassée correctly perceived that engineers like to work on hot products, but he made the mistake of defining "hot" as "high end," dooming Apple's efforts in the home and small business markets.

Gassée's organization was filled with meek sycophants. In his staff meetings, Jean-Louis talked, and everyone else listened. There was no healthy discussion, no wild and crazy brainstorming that Apple had been known for and that had produced the company's most innovative programs. It was like Stalin's staff meeting.

Another early Sculley favorite was Allen Loren, who came to Apple as head of management information systems—the chief administrative computer guy—and then suddenly found himself in charge of sales and marketing simply because Sculley liked him. Loren was a good MIS guy but a bad marketing and sales guy.

Loren presided over Apple's single greatest disaster, the

price increase of 1988. In an industry built around the concept of prices' continually dropping, Loren decided to raise prices on October 1, 1988, in an effort to raise Apple's sinking profit margins. By raising prices Loren was fighting a force of nature, like asking the earth to reverse its direction of rotation, the tides to stop, mothers everywhere to stop telling their sons to get haircuts. Ignorantly, he asked the impossible, and the bottom dropped out of Apple's market. Sales tumbled, market share tumbled. Any momentum that Apple had was lost, maybe for years, and Sculley allowed that to happen.

Loren was followed as vice-president of marketing by David Hancock, who was known throughout Apple as a blowhard. When Apple marketing should have been trying to recover from Loren's pricing mistake, the department did little under Hancock. The marketing department was instead distracted by nine reorganizations in less than two years. People were so busy covering their asses that they weren't working, so Apple's business in 1989 and 1990 showed what happens when there is no marketing at all.

The whole marketing operation at Apple is now run by former salespeople, a dangerous trend. Marketing is the creation of long-term demand, while sales is execution of marketing strategies. Marketing is buying the land, choosing what crop to grow, planting the crop, fertilizing it, and then deciding when to harvest. Sales is harvesting the crop. Salespeople in general don't think strategically about the business, and it's this short-term focus that's prevalent right now at Apple.

When Apple introduced its family of lower-cost Macintoshes in the fall of 1990, marketing was totally unprepared for their popularity. The computer press had been calling for lower-priced Macs, but nobody inside Apple expected to sell a lot of the boxes. Blame this on the lack of marketing, and also blame it on the demise, two years before, of Apple's entire market research department, which fell in another political game. When the Macintosh Classic, LC, and IIsi appeared, their overwhelming popularity surprised, pleased, but then dismayed Apple, which was still staffing up as a company that sold expensive computers. Profit margins dropped despite an 85 percent increase in sales, and Sculley found himself having to lay off 15 percent of Apple's work force, because of unexpected success that should have been, could have been, planned for.

Sculley's current favorite is Fred Forsythe, formerly head of manufacturing but now head of engineering, with major responsibility for research and development. Like Loren, Forsythe was good at the job he was originally hired to do, but that does not at all mean he's the right man for the R&D job. Nor is Sculley, who has taken to calling himself Apple's Chief Technical Officer—an insult to the company's real engineers.

So why does Sculley make these terrible personnel moves? Maybe he wants to make sure that people in positions of power are loyal to him, as all these characters are. And by putting them in jobs they are not really up to doing, they are kept so busy that there is no time or opportunity to plot against Sculley. It's a stupid reason, I know, and one that has cost Apple billions of dollars, but it's the only one that makes any sense.

With all the ebb and flow of people into and out of top management positions at Apple, it reached the point where it was hard to get qualified people even to accept top positions, since they knew they were likely to be fired. That's when Sculley started offering signing bonuses. Joe Graziano, who'd left Apple to be the chief financial officer at Sun Microsystems, was lured back with a $1.5 million bonus in 1990. Shareholders and Apple employees who weren't raking in such big rewards complained about the bonuses, but the truth is that it was the only way Sculley could get good people to work for him. (Other large sums are often counted in "Graz" units. A million and a half dollars is now known as "1 Graz"—a large unit of currency in Applespeak.)

The rest of the company was as confused as its leadership. Somehow, early on, reorganizations—"reorgs"—became part of the Apple culture. They happen every three to six months and come from Apple's basic lack of understanding that people need stability in order to be able to work together.

Reorganizations have become so much of a staple at Apple that employees categorize them into two types. There's the "Flint Center reorganization," which is so comprehensive that Apple calls its Cupertino workers into the Flint Center auditorium at DeAnza College to hear the top executives explain it. And there's the smaller "lunchroom reorganization," where Apple managers call a few departments into a company cafeteria to hear the news.

The problem with reorgs is that they seem to happen overnight, and many times they are handled by groups being demolished and people being told to go to Human Resources and find a new job at Apple. And so the sense is at Apple that if you don't like where you are, don't worry, because three to six months from now everything is going to be different. At the same time, though, the continual reorganizations mean that nobody has long-term responsibility for *anything*. Make a bad decision? Who cares! By the time the bad news arrives, you'll be gone and someone else will have to handle the problems.

If you do like your job at Apple, watch it, because unless you are in some backwater that no one cares about and is severely understaffed, your job may be gone in a second, and you may be "on the street," with one or two months to find a job at Apple.

Today, the sense of anomie—alienation, disconnectedness—at Apple is major. The difference between the old Apple, which was crazy, and the new Apple is anomie. People are alienated. Apple still gets the bright young people. They come into

Apple, and instead of getting all fired up about something, they go through one or two reorgs and get disoriented. I don't hear people who are really happy to be at Apple anymore. They wonder why they are there, because they've had two bosses in six months, and their job has changed twice. It's easy to mix up groups and end up not knowing anyone. That's a real problem.

"I don't know what will happen with Apple in the long term," said Larry Tesler. "It all depends on what they do."

They? Don't you mean *we,* Larry? Has it reached the point where an Apple vice-president no longer feels connected to his own company?

With the company in a constant state of reorganization, there is little sense of an enduring commitment to strategy at Apple. It's just not in the culture. Surprisingly, the company *has* a commitment to doing good products; it's the follow-through that suffers. Apple specializes in flashy product introductions but then finds itself wandering away in a few weeks or months toward yet another pivotal strategy and then another.

Compare this with Microsoft, which is just the opposite, doing terrific implementation of mediocre products. For example, in the area of multimedia computing—the hot new product classification that integrates computer text, graphics, sound, and full-motion video—Microsoft's Multimedia Windows product is ho-hum technology acquired from a variety of sources and not very well integrated, but the company has implemented it very well. Microsoft does a good roll-out, offers good developer support, and has the same people leading the operation for years and years. They follow the philosophy that as long as you are the market leader and are still throwing technology out there, you won't be dislodged.

Microsoft is taking the Japanese approach of not caring how long or how much money it takes to get multimedia right. They've been at it for six years so far, and if it takes another six years, so be it. That's what makes me believe Microsoft will continue to be a factor in multimedia, no matter how bad its products are.

In contrast to Microsoft, Apple has a very elegant multimedia architecture called QuickTime, which does for time-based media what Apple's QuickDraw did for graphics. QuickTime has tools for integrating video, animation, and sound into Macintosh programs. It automatically synchronizes sound and images and provides controls for playing, stopping, and editing video sequences. QuickTime includes technology for compressing images so they require far less memory for storage. In short, QuickTime beats the shit out of Microsoft's Multimedia Extensions for Windows, but Apple is also taking a typical short-term view. Apple produced a flashy intro, but has no sense of enduring commitment to its own strategy.

The good and the bad that was Apple all came from Steve Jobs, who in 1985 was once again an orphan and went off to found another company—NeXT Inc.—and take another crack at playing the father role. Steve sold his Apple stock in a huff (and at a stupidly low price), determined to do it all over again—to build another major computer company—and to do it his way.

"Steve never knew his parents," recalled Trip Hawkins, who went to Apple as manager of market planning in 1979. "He makes so much noise in life, he cries so loud about everything, that I keep thinking he feels that if he just cries loud enough, his real parents will hear and know that they made a mistake giving him up."

FONT WARS

Of the 5 billion people in the world, there are only four who I'm pretty sure have stayed consistently on the good side of Steve Jobs. Three of them—Bill Atkinson, Rich Page, and Bud Tribble—all worked with Jobs at Apple Computer. Atkinson and Tribble are code gods, and Page is a hardware god. Page and Tribble left Apple with Jobs in 1985 to found NeXT Inc., their follow-on computer company, where they remain in charge of hardware and software development, respectively.

So how did Atkinson, Page, and Tribble get off so easily when the rest of us have to suffer through the rhythmic pattern of being ignored, then seduced, then scourged by Jobs? Simple; among the three, they have the total brainpower of a typical Third World country, which is more than enough to make even Steve Jobs realize that he is, in comparison, a single-celled, carbon-based life form. Atkinson, Page, and Tribble have answers to questions that Jobs doesn't even know he should ask.

The fourth person who has remained a Steve Jobs favorite is John Warnock, founder of Adobe Systems. Warnock is the father that Steve Jobs always wished for. He's also the man who made possible the Apple LaserWriter printer and desktop publishing. He's the man who saved the Macintosh.

Warnock, one of the world's great programmers, has the technical ability that Jobs lacks. He has the tweedy, professorial style of a Robert Young, clearly contrasting with the blue-collar vibes of Paul Jobs, Steve's adoptive father. Warnock has a passion, too, about just the sort of style issues that are so important to Jobs. Warnock is passionate about the way words and pictures look on a computer screen or on a printed page, and Jobs respects that passion.

Both men are similar, too, in their unwillingness to compromise. They share a disdain for customers based on their conviction that the customer can't even imagine what they (Steve and John) know. The customer is so primitive that he or she is not even qualified to say what they need.

Welcome to the Adobe Zone.

John Warnock's rise to programming stardom is the computer science equivalent of Lana Turner's being discovered sitting in Schwab's Drugstore in Hollywood. He was a star overnight.

A programmer's life is spent implementing algorithms, which are just specific ways of getting things done in a computer program. Like chess, where you may have a Finkelstein opening or a Blumberg entrapment, most of what a programmer does is fitting other people's algorithms to the local situation. But every good programmer has an algorithm or two that is all his or hers, and most programmers dream of that moment when they'll see more clearly than they ever have before the answer to some incredibly complex programming problem, and their particular solution will be added to the algorithmic lore of programming. During their fifteen minutes of techno-fame, everyone who is anyone in the programming world will talk about the Clingenpeel shuffle or the Malcolm X sort.

Most programmers don't ever get that kind of instant glory, of course, but John Warnock did. Warnock's chance came when he was a graduate student in mathematics, working at the University of Utah computer center, writing a mainframe program to automate class registration. It was a big, dumb program, and Warnock, who like every other man in Utah had a wife and kids to support, was doing it strictly for the money.

Then Warnock's mindless toil at the computer center was interrupted by a student who was working on a much more challenging problem. He was trying to write a graphics program to present on a video monitor an image of New York harbor as seen from the bridge of a ship. The program was supposed to run in real time, which meant that the video ship would be moving in the harbor, with the view slowly shifting as the ship changed position.

The student was stumped by the problem of how to handle the view when one object moved in front of another. Say the video ship was sailing past the Statue of Liberty, and behind the statue was the New York skyline. As the ship moved forward, the buildings on the skyline should appear to shift behind the statue, and the program would have to decide which parts of the buildings were blocked by the statue and find a way to turn off just those parts of the image, shaping the region of turned-off image to fit along the irregular profile of the statue. Put together dozens of objects at varying distances, all shifting in front of or behind each other, and just the calculation of what could and couldn't be visible was bringing the computer to its knees.

"Why not do it this way?" Warnock asked, looking up from his class registration code and describing a way of solving the problem that had never been thought of before, a way so simple that it should have been obvious but had somehow gone unthought of by the brightest programming minds at the university. No big deal.

Except that it *was* a big deal. Dumbfounded by Warnock's

casual brilliance, the student told his professor, who told the department chairman, who told the university president, who must have told God (this is Utah, remember), because the next thing he knew, Warnock was giving talks all over the country, describing how he solved the hidden surface problem. The class registration program was forever forgotten.

Warnock switched his Ph.D. studies from mathematics to computer science, where the action was, and was soon one of the world's experts on computer graphics.

Computer graphics, the drawing of pictures on-screen and on-page, is very difficult stuff. It's no accident that more than 80 percent of each human brain is devoted to processing visual data. Looking at a picture and deciding what it portrays is a major effort for humans, and often an impossible one for computers.

Jump back to that image of New York harbor, which was to be part of a ship's pilot training simulator ordered by the U.S. Maritime Academy. How do you store a three-dimensional picture of New York harbor inside a computer? One way would be to put a video camera in each window of a real ship and then sail that ship everywhere in the harbor to capture a video record of every vista. This would take months, of course, and it wouldn't take into account changing weather or other ships moving around the harbor, but it would be a start. All the video images could then be digitized and stored in the computer. Deciding what view to display through each video window on the simulator would be just a matter of determining where the ship was supposed to be in the harbor and what direction it was facing, and then finding the appropriate video scene and displaying it. Easy, eh? But how much data storage would it require?

Taking the low-buck route, we'll require that the view only be in typical PC resolution of 640-by-400 picture elements (pixels), which means that each stored screen will hold 256,000 pixels.

Since this is 8-bit color (8 bits per pixel), that means we'll need 256,000 bytes of storage (8 bits make 1 byte) for each screen image. Accepting a certain jerkiness of apparent motion, we'll need to capture images for the video database every ten feet, and at each of those points we'll have to take a picture in at least eight different directions. That means that for every point in the harbor, we'll need 2,048,000 bytes of storage. Still not too bad, but how many such picture points are there in New York harbor if we space them every ten feet? The harbor covers about 100 square miles, which works out to 27,878,400 points. So we'll need just over 57 billion bytes of storage to represent New York harbor in this manner. Twenty years ago, when this exercise was going on in Utah, there was no computer storage system that could hold 57 billion bytes of data or even 5.7 billion bytes. It was impossible. And the system would have been terrifically limited in other ways, too. What would the view be like from the top of the Statue of Liberty? Don't know. With all the data gathered at sea level, there is no way of knowing how the view would look from a higher altitude.

The problem with this type of computer graphics system is that all we are doing is storing and calling up bits of data rather than twiddling them, as we should do. Computers are best used for processing data, not just retrieving them. That's how Warnock and his buddies in Utah solved the data storage problem in their model of New York harbor. Rather than take pictures of the whole harbor, they *described* it to the computer.

Most of New York harbor is empty water. Water is generally flat with a few small waves, it's blue, and it lives its life at sea level. There I just described most of New York harbor in eighteen words, saving us at least 50 billion bytes of storage. What we're building here is an imaging model, and it assumes that the default appearance of New York harbor is wet. Where it's not wet—where there are piers or buildings or islands—I can describe those, too, by telling the computer what the object looks like and where it is positioned in space. What I'm actually doing is telling the computer how to draw a picture of the object, specifying characteristics like size, shape, and color. And if I've already described a tugboat, for example, and there are dozens of tugboats in the harbor that look alike, the next time I need to describe one I can just refer back to the earlier description, saying to draw another tugboat and another and another, with no additional storage required.

This is the stuff that John Warnock thought about in Utah and later at Xerox PARC, where he and Martin Newell wrote a language they called JaM, for *John and Martin*. JaM provided a vocabulary for describing objects and positioning them in a three-dimensional database. JaM evolved into another language called Interpress, which was used to describe words and pictures to Xerox laser printers. When Warnock was on his own, after leaving Xerox, Interpress evolved into a language called PostScript. JaM, Interpress, and PostScript are really the same language, in fact, but for reasons having to do with copyrights and millions of dollars, we pretend that they are different.

In PostScript, the language we'll be talking about from now on, there is no difference between a tugboat or the letter E. That is, PostScript can be used to draw pictures of tugboats and pictures of the letter E, and to the PostScript language each is just a picture. There is no cultural or linguistic symbolism attached to the letter, which is, after all, just a group of straight and curved lines filled in with color.

PostScript describes letters and numbers as mathematical formulas rather than as bit maps, which are just patterns of tiny dots on a page or screen. PostScript popularized the outline font, where a description of each letter is stored as a formula for lines and bezier curves and recipes for which parts of the character are to be filled with color and which parts are not. Outline fonts, because they are based on mathematical descriptions of each letter, are resolution independent; they can be scaled up or

down in size and printed in as fine detail as the printer or typesetter is capable of producing. And like the image of a tugboat,
which increases in detail as it sails closer, PostScript outline fonts
contain "hints" that control how much detail is given up as type
sizes get smaller, making smaller type sizes more readable than
they otherwise would be.

Before outline fonts can be printed, they have to be raster-
ized, which means that a description of which bits to print where
on the page has to be generated. Before there were outline fonts,
bit-mapped fonts were all there were, and they were generated in
a few specific sizes by people called fontographers, not computers.
But with PostScript and outline fonts, it's as easy to generate a
10.5-point letter as the usual 10-, 12-, or 14-point versions.

Warnock and his boss at Xerox, Chuck Geschke, tried for two
years to get Xerox to turn Interpress into a commercial product.
Then they decided to start their own company with the idea of
building the most powerful printer in history, to which people
would bring their work to be beautifully printed. Just as Big Blue
imagined there was a market for only fifty IBM 650 mainframes,
the two ex-Xerox guys thought the world needed only a few Post-
Script printers.

Warnock and Geschke soon learned that venture capitalists
don't like to fund service businesses, so they next looked into creating a computer workstation with custom document preparation
software that could be hooked into laser printers and typesetters,
to be sold to typesetting firms and the printing departments of
major corporations. Three months into *that* business, they discovered at least four competitors were already underway with similar
plans and more money. They changed course yet again and
became sellers of graphics systems software to computer companies, designers of printer controllers featuring their PostScript language, and the first seller of PostScript fonts.

Adobe Systems was named after the creek that ran past

Warnock's garden in Los Altos, California. The new company defined the PostScript language and then began designing printer controllers that could interpret PostScript commands, rasterize the image, and direct a laser engine to print it on page. That's about the time that Steve Jobs came along.

The usual rule is that hardware has to exist before programmers will write software to run on it. There are a few exceptions to this rule, and one of these is PostScript, which is very advanced, very complex software that *still* doesn't run very fast on today's personal computers. PostScript was an order of magnitude more complex than most personal computer software of the mid-1980s. Tim Paterson's Quick and Dirty Operating System was written in less than six months. Jonathan Sachs did 1-2-3 in a year. Paul Allen and Bill Gates pulled together Microsoft BASIC in six weeks. Even Andy Hertzfeld put less than two years into writing the system software for Macintosh. But PostScript took twenty man-years to perfect. It was the most advanced software ever to run on a personal computer, and few microcomputers were up to the task.

The mainframe world, with its greater computing horsepower, might logically have embraced PostScript printers, so the fact that the personal computer was where PostScript made its mark is amazing, and is yet another testament to Steve Jobs's will.

The 128K Macintosh was a failure. It was an amazing design exercise that sat on a desk and did next to nothing, so not many people bought early Macs. The mood in Cupertino back in 1984 was gloomy. The Apple III, the Lisa, and now the Macintosh were all failures. The Apple II division was being ignored, the Lisa division was deliberately destroyed in a fit of Jobsian pique, and the Macintosh division was exhausted and depressed.

Apple had $250 million sunk in the ground before it started

making money on the Macintosh. Not even the enthusiasm of Steve Jobs could make the world see a 128K Mac with a floppy disk drive, two applications, and a dot-matrix printer as a viable business computer system.

Apple employees may drink poisoned Kool-Aid, but Apple customers don't.

It was soon evident, even to Jobs, that the Macintosh needed a memory boost and a compelling application if it was going to succeed. The memory boost was easy, since Apple engineers had secretly included the ability to expand memory from 128K to 512K, in direct defiance of orders from Jobs. Coming up with the compelling application was harder; it demanded patience, which was never seen as a virtue at Apple.

The application so useful that it compels people to buy a specific computer doesn't have to be a spreadsheet, though that's what it turned out to be for the Apple II and the IBM PC. Jobs and Sculley thought it would be a spreadsheet, too, that would spur sales of the Mac. They had high hopes for Lotus Jazz, which turned up too late and too slow to be a major factor in the market. There was, as always, a version of Microsoft's Multiplan for the Mac, but that didn't take off in the market either, primarily because the Mac, with its small screen and relatively high price, didn't offer a superior environment for spreadsheet users. For running spreadsheets, at least, PCs were cheaper and had bigger screens, which was all that really mattered.

For the Lisa, Apple had developed its own applications, figuring that the public would latch onto one of the seven as the compelling application. But while the Macintosh came with two bundled applications of its own—MacWrite and MacPaint—Jobs wanted to do things in as un-Lisa-like manner as possible, which meant that the compelling application would have to come from outside Apple.

Mike Boich was put in charge of what became Apple's

Macintosh evangelism program. Evangelists like Alain Rossmann and Guy Kawasaki were sent out to bring the word of Macintosh to independent software developers, giving them free computers and technical support. They hoped that these efforts would produce the critical mass of applications needed for the Mac to survive and at least one compelling application that was needed for the Mac to succeed.

There are lots of different personal computers in the world, and they all need software. But little software companies, which describes about 90 percent of the personal computer software companies around, can't afford to make too many mistakes by developing applications for computers that fail in the marketplace. At Electronic Arts, Trip Hawkins claims to have been approached to develop software for sixty different computer types over six or seven years. Hawkins took a chance on eighteen of those systems, while most companies pick only one or two.

When considering whether to develop for a different computer platform, software companies are swayed by an installed base—the number of computers of a given type that are already working in the world—by money, and by fear of being left behind technically. Boich, Rossmann, and Kawasaki had no installed base of Macintoshes to point to. They couldn't claim that there were a million or 10 million Macintoshes in the world, with owners eager to buy new and innovative applications. And they didn't have money to pay developers to do Mac applications—something that Hewlett-Packard and IBM had done in the past.

The pitch that worked for the Apple evangelists was to cultivate the developers' fear of falling behind technically. "Graphical user interfaces are the future of computing," they'd say, "and this is the best graphical user interface on the market right now. If you aren't developing for the Macintosh, five years from now your company won't be in business, no matter what graphical platform is dominant then."

The argument worked, and 350 Macintosh applications were soon under development. But Apple still needed new technology that would set the Mac apart from its graphical competitors. The Lisa and the Xerox Star had not been ignored by Apple's competitors, and a number of other graphical computing environments were announced in 1983, even before the Macintosh shipped.

VisiCorp was betting (and losing) its corporate existence on a proprietary graphical user interface and software for IBM PCs and clones called VisiOn. VisiOn appeared in November 1983, more than a year after it was announced. With VisiOn, you got a mouse, a special circuit card that was installed inside the PC, and software including three applications—word processing, spreadsheet, and graphics. VisiOn offered no color, no icons, and it was slow—all for a list price of $1,795. The shipping version was supposed to have been twelve times faster than the demo; it wasn't. Developers hated VisiOn because they had to pay a big up-front fee to get the information needed to write programs (literally *anti*-evangelism) and then had to buy time on a Prime minicomputer, the only computer environment in which applications could be developed. VisiOn was a dud, but until it was actually out, failing in the world, it had a lot of people scared.

One person who was definitely scared by VisiOn was Bill Gates of Microsoft, who stood transfixed through three complete VisiOn demonstrations at the Comdex computer trade show in 1982. Gates had Charles Simonyi fly down from Seattle just to see the VisiOn demo, then Gates immediately went back to Bellevue and started his own project to throw a graphical user interface on top of DOS. This was the Interface Manager, later called Microsoft Windows, which was announced in 1983 and shipped in 1985. Windows was slow, too, and there weren't very many applications that supported the environment, but it fulfilled Gates' goal, which was not to be the best graphical environment around, but simply to defend the DOS franchise. If the world wanted a graphical user interface, Gates would add one to DOS. If they want a pen-based interface, he'll add one to DOS (it's called Windows for Pen Computing). If the world wants voice recognition, or multimedia, or fingerpainting input, Gates will add it to DOS, because DOS, and the regular income it provides, year after year, funds everything else at Microsoft. DOS *is* Microsoft.

Gates did Windows as a preemptive strike against VisiOn, and he developed Microsoft applications for the Macintosh, because it was clear that Windows would not be good enough to stop the Mac from becoming a success. Since he couldn't beat the Macintosh, Gates supported it, and in turn gained knowledge of graphical environments. He also made an agreement with Apple allowing him to use certain Macintosh features in Windows, an agreement that later landed both companies in court.

Finally, there was GEM, another graphical environment for the IBM PC, which appeared from Gary Kildall's Digital Research, also in 1983. GEM is still out there, in fact, but the only GEM application of note is Ventura Publisher, a popular desktop publishing package for the IBM world, ironically sold by Xerox. Most Ventura users don't even know they are using GEM.

Apple needed an edge against all these would-be competitors, and that edge was the laser printer. Hewlett-Packard introduced its LaserJet printer in 1984, setting a new standard for PC printing, but Steve Jobs wanted something much, much better, and when he saw the work that Warnock and Geschke were doing at Adobe, he knew they could give him the sort of printer he wanted. H-P's LaserJet output looked as if it came from a typewriter, while Jobs was determined that his LaserWriter output would look like it came from a type*setter*.

Jobs used $2.5 million to buy 15 percent of Adobe, an

extravagant move that was wildly unpopular among Apple's top management, who generally gave up the money for lost and moved to keep Jobs from making other such investments in the future. Apple's investment in Adobe was far from lost though. It eventually generated more than $10 billion in sales for Apple, and the stock was sold six years later for $89 million. Still, in 1984, conventional wisdom said the Adobe investment looked like a bad move.

The Apple LaserWriter used the same laser print mechanism that H-P's LaserJet did. It also used a special controller card that placed inside the printer what was then Apple's most powerful computer; the printer itself was a computer. Adobe designed a printer controller for the LaserWriter, and Apple designed one too. Jobs arrogantly claimed that nobody—not even Adobe—could engineer as well as Apple, so he chose to use the Apple-designed controller. For many years, this was the only non-Adobe-designed PostScript controller on the market. The first generation of competitive PostScript printers from other companies all used the rejected Adobe controller and were substantially faster as a result.

The LaserWriter cost $7,000, too much for a printer that would be available to only a single microcomputer. Jobs, who still didn't think that workers needed umbilical cords to their companies, saw the logic in at least having an umbilical cord to the LaserWriter, and so AppleTalk was born. AppleTalk was clever software that worked with the Zilog chip that controlled the Macintosh serial port, turning it into a medium-speed network connection. AppleTalk allowed up to thirty-two Macs to share a single LaserWriter.

At the same time that he was ordering AppleTalk, Jobs still didn't understand the need to link computers together to share information. This antinetwork bias, which was based on his concept of the lone computist—a digital Clint Eastwood character who, like Jobs, thought he needed nobody else—persisted even years later when the NeXT computer system was introduced in 1988. Though the NeXT had built-in Ethernet networking, Jobs was still insisting that the proper use of his computer was to transfer data on a removable disk. He felt so strongly about this that for the first year, he refused orders for NeXT computers that were specifically configured to store data for other computers on the network. That would have been an impure use of his machine.

Adobe Systems rode fonts and printer software to more than $100 million in annual sales. By the time they reach that sales level, most software companies are being run by marketers rather than by programmers. The only two exceptions to this rule that I know of are Microsoft and Adobe—companies that are more alike than their founders would like to believe.

Both Microsoft and Adobe think they are following the organizational model devised by Bob Taylor at Xerox PARC. But where Microsoft has a balkanized version of the Taylor model, got second-hand through Charles Simonyi, Warnock and Geschke got their inspiration directly from the master himself. Adobe is the closest a commercial software company can come to following Taylor's organizational model and still make a profit.

The problem, of course, is that Bob Taylor's model isn't a very good one for making products *or* profits—it was never intended to be—and Adobe has been able to do both only through extraordinary acts of will.

As it was at PARC, what matters at Adobe is technology, not marketing. The people who matter are programmers, not marketers. Ideologically correct technology is more important than making money—a philosophy that clearly differentiates Adobe from Microsoft, where making money is the prime directive.

John Warnock looks at Microsoft and sees only shoddy technology. Bill Gates looks at Adobe and sees PostScript monks who are ignoring the real world—the world controlled by Bill Gates. And it's true; the people of Adobe see PostScript as a religion and hate Gates because he doesn't buy into that religion.

There is a part of John Warnock that would like to have the same fatherly relationship with Bill Gates that he already has with Steve Jobs. But their values are too far apart, and, unlike Steve, Bill already has a father.

Being technologically correct is more important to Adobe than pleasing customers. In fact, pleasing customers is relatively unimportant. Early in 1985, for example, representatives from Apple came to ask Adobe's help in making the Macintosh's bit-mapped fonts print faster. These were programmers from Adobe's largest customer who had swallowed their pride to ask for help. Adobe said, "No."

"They wanted to dump screens [to the printer] faster, and they wanted Apple-specific features added to the printer," Warnock explained to me years later. "Apple came to me and said, 'We want you to extend PostScript in a way that is proprietary to Apple.' I had to say no. What they asked would have destroyed the value of the PostScript standard in the long term."

If a customer that represented 75 percent of my income asked me to walk his dog, wash her car, teach their kids to read, or to help find a faster way to print bit-mapped fonts, I'd do it, even if it meant adding a couple proprietary features to PostScript, which already had lots of proprietary features—proprietary to Adobe.

The scene with Apple was quickly forgotten, because putting bad experiences out of mind is the Adobe way. Adobe is like a family that pretends grandpa isn't an alcoholic. Unlike Microsoft, with its screaming and willingness to occasionally ship schlock code, all that matters at Adobe is great technology and the appearance of calm.

A Stanford M.B.A. was hired to work as Adobe's first evangelist, trying to get independent software developers to write PostScript applications. Technical evangelism usually means going on the road—making contacts, distributing information, pushing the product. Adobe's evangelist went more than a year without leaving the building on business. He spent his days up in the lab, playing with the programmers. His definition of evangelism was waiting for potential developers to call him, *if* they knew he existed at all. What's amazing about this story is that this nonevangelist came under no criticism for his behavior. Nobody said a thing.

Nobody said anything, too, when a technical support worker occasionally appeared at work wearing a skirt. Nobody said, "Interesting skirt, Glenn." Nobody said anything.

Some folks from Adobe came to visit *InfoWorld* one afternoon, and I asked about Display PostScript, a product that had been developed to bring PostScript fonts and graphics to Macintosh screens. Display PostScript had been licensed to Aldus for a new version of its PageMaker desktop publishing program called PageMaker Pro. But at the last minute, after the product was finished and the deal with Aldus was signed, Adobe decided that it didn't want to do Display PostScript for the Macintosh after all. They took the product back, and scrambled hard to get Aldus to cancel PageMaker Pro, too. I wanted to know why they withdrew the product.

The product marketing manager for PostScript, the person whose sole function was to think about how to get people to buy more PostScript, claimed to have never heard of Display PostScript for the Mac or of PageMaker Pro. He looked bewildered.

"That was before you joined the company," explained Steve MacDonald, an Adobe vice-president who was leading the group.

"You don't tell new marketing people the history of their own products?" I asked, incredulous. "Or is it just the mistakes you don't tell them about?"

MacDonald shrugged.

For all its apparent disdain for money, Adobe has an incredible ability to wring the stuff out of customers. In 1989, for example, every Adobe programmer, marketing executive, receptionist, and shipping clerk represented $357,000 in sales and $142,000 in profit. Adobe has the highest profit margins and the greatest sales per employee of any major computer hardware or software company, but such performance comes at a cost. Under the continual prodding of the company's first chairman, a venture capitalist named Q. T. Wiles, Adobe worked hard to maximize earnings per share, which boosted the stock price. Warnock and Geschke, who didn't know any better, did as Q. T. told them to.

Q. T. is gone now, his Adobe shares sold, but the company is trapped by its own profitability. Earnings per share are supposed to only rise at successful companies. If you earned a dollar per share last year, you had better earn $1.20 per share this year. But Adobe, where 400 people are responsible for more than $150 million in sales, was stretched thin from the start. The only way that the company could keep its earnings going ever upward was to get more work out of the same employees, which means that the couple of dozen programmers who work most of the technical miracles are under terrific pressure to produce.

This pressure to produce first became a problem when Warnock decided to do Adobe Illustrator, a PostScript drawing program for the Macintosh. Adobe's customers to that point were companies like Apple and IBM, but Illustrator was meant to be sold to you and me, which meant that Adobe suddenly needed distributors, dealers, printers for manuals, duplicators for floppy disks—things that weren't at all necessary when serving customers meant sending a reel of computer tape over to Cupertino in exchange for a few million dollars, thank you. But John Warnock wanted the world to have a PostScript drawing tool, and so the world would have a PostScript drawing tool. A brilliant programmer named Mike Schuster was pulled away from the company's system software business to write the application as Warnock envisioned it.

In the retail software business, you introduce a product and then immediately start doing revisions to stay current with technology and fix bugs. John Warnock didn't know this. Adobe Illustrator appeared in 1986, and Schuster was sent to work on other things. They should have kept someone working on Illustrator, improving it and fixing bugs, but there just wasn't enough spare programmer power to allow that. A version of Illustrator for the IBM PC followed that was so bad it came to be called the "landfill version" inside the company. PC Illustrator should have been revised instantly, but wasn't.

When Adobe finally got around to sprucing up the Macintosh version of Illustrator, they cleverly called the new version Illustrator 88, because it appeared in 1988. You could still buy Illustrator 88 in 1989, though. And in 1990. And even into 1991, when it was finally replaced by Illustrator 3.0. Adobe is not a marketing company.

In 1988, Bill Gates asked John Warnock for PostScript code and fonts to be included with the next version of Windows. With Adobe's help users would be able to see the same beautiful printing on-screen that they could print on a PostScript printer. Gates, who never pays for anything if he can avoid it, wanted the code for free. He argued that giving PostScript code to Microsoft would lead to a dramatic increase in Adobe's business selling fonts, and Adobe would benefit overall. Warnock said, "No."

In September 1989, Apple Computer and Microsoft an-

nounced a strategic alliance against Adobe. As far as both companies were concerned, John Warnock had said "No" twice too
often. Apple was giving Microsoft its software for building fonts
in exchange for use of a PostScript clone that Microsoft had
bought from a developer named Cal Bauer.

Forty million Apple dollars were going to Adobe each year,
and clever Apple programmers, who still remembered being rejected by Adobe in 1985, were arguing that it would be cheaper to
roll their own printing technology than to continue buying Adobes.

In mid-April, news had reached Adobe that Apple would
soon announce the phasing out of PostScript in favor of its own
code, to be included in the upcoming release of new Macintosh
control software called System 7.0. A way had to be found fast to
counter Apple's strategy or change it.

Only a few weeks after learning Apple's decision—and before anything had been announced by Apple or Microsoft—
Adobe Type Manager, or ATM, was announced—software that
would bring Adobe fonts directly to Macintosh screens without
the assistance of Apple since it would be sold directly to users.
ATM, which would work only with fonts—with words rather
than pictures—was replacing Display PostScript, which Adobe
had already tried (and failed) to sell to Apple. ATM had the
advantage over Apple's System 7.0 software that it would work
with older Macintoshes. Adobe's underlying hope was that quick
market acceptance of ATM would dissuade Apple from even setting out on its separate course.

But Apple made its announcement anyway, sold all its
Adobe shares, and joined forces with Microsoft to destroy its former ally. Adobe's threat to both Apple and Microsoft was so great
that the two companies conveniently ignored their own year-
long court battle over the vestiges of an earlier agreement
allowing Microsoft to use the look and feel of Apple's Macintosh
computer in Microsoft Windows.

Apple-Microsoft and Apple-Adobe are examples of strategic alliances as they are conducted in the personal computer industry. Like bears mating or teenage romances, strategic alliances are important but fleeting.

Apple chose to be associated with Adobe only as long as the relationship worked to Apple's advantage. No sticking with old friends through thick and thin here.

For Microsoft, fonts and printing technology had been of little interest, since Gates saw as important what happened inside the box, not inside the printer. Then IBM decided it wanted the same fonts in both its computers and printers, only to discover that Microsoft, its traditional software development partner, had no font technology to offer. So IBM began working with Adobe and listening to the ideas of John Warnock.

If IBM is God in the PC universe then Bill Gates is the pope. Warnock, now talking directly with IBM, was both a heretic and a threat to Gates. Warnock claimed that Gates was not a good servant of God, that Microsoft's technology was inferior. Worse, Warnock was correct, and Gates knew it. Control of the universe in the box was at stake.

Warnock and Adobe had to die, Gates decided, and if it took an unholy alliance with Apple and a temporary putting aside of legal conflicts between Microsoft and Apple to kill Adobe, then so be it.

This passion play of Adobe, Apple, and Microsoft could have taken place between companies in many industries, but what sets the personal computer industry apart is that the products in question—Adobe Type Manager and Apple's System 7.0—did not even exist.

Battles of midsized cars or two-ply toilet tissue take place on showroom floors and supermarket shelves, but in the personal computer industry, deals are cut and share prices fluctuate on the supposed attributes of products that have yet to be written or even fully designed. Apple's offensive against Adobe was based on revealing the ongoing development of software that users could not expect to purchase for at least a year (two years, it turned out); Adobe's response was a program that would take months to develop.

ATM was announced, *then* developed, essentially by a single programmer who used to joke with the Adobe marketing manager about whether the product or its introduction would be done first.

Both companies were dueling with intentions, backed up by the conviction of some computer hacker that given enough time and junk food, he could eventually write software that looked pretty much like what had just been announced with such fanfare.

As I said, computer graphics software is *very* hard to do well. By the middle of 1991, Apple and Adobe had made friends again, in part because Microsoft had not been able to fulfill its part of the deal with Apple. "Our entry into the printer software business has not succeeded," Bill Gates wrote in a memo to his top managers. "Offering a cheap PostScript clone turned out to not only be very hard but completely irrelevant to helping our other problems. We overestimated the threat of Adobe as a competitor and ended up making them an 'enemy,' while we hurt our relationship with Hewlett-Packard . . . "

*Over*estimated the threat of Adobe as a competitor? In a way it's true, because the computer world is moving on to other issues, leaving Adobe behind. Adobe makes more money than ever in its PostScript backwater, but is not wresting the operating system business from Microsoft, as both companies had expected.

With its reliance on only a few very good programmers. Adobe was forced to defend its existing businesses at the cost of its future. John Warnock is still a better programmer than Bill Gates, but he'll never be as savvy.

CHAPTER TWELVE

ON THE BEACH

America's advantage in the PC business doesn't come from our education system, from our fluoridated water, or, Lord knows, from our tax structure. And it doesn't come from some innate ability we have to run big companies with thousands of employees and billions in sales. The main thing America has had going for it is the high-tech start-up, and, of course, our incredible willingness to fail.

One winter back at the College of Wooster, in Wooster, Ohio, I took a bowling course that changed my life. P.E. courses were mandatory, and the only alternative that quarter, as I remember it, was a class in snow shoveling.

A dozen of us met in the bowling alley three times a week for ten weeks. The class was about evenly divided between men and women, and all we had to do was show up and bowl, handing in our score sheets at the end of each session to prove we'd been there. I remember bowling a 74 in that first game, but my scores quickly improved with practice. By the fourth week, I'd stabilized in the 140–150 range and didn't improve much after that.

Four of us always bowled together: my roommate, two women of mystery (all women were women of mystery to me then), and me. My roommate, Bob Scranton, was a better bowler than I was, and his average settled in the 160–170 range at mid-term. But the two women, who started out bowling scores in the 60s, improved steadily over the whole term, adding a few points each week to their averages, peaking in the tenth week at around 140.

When our grades appeared, the other Bob and I got Bs, and the two women of mystery received As.

"Don't you understand?" one of the women tried to explain. "They grade on improvement, so all we did was make sure that our scores got a little better each week, that's all."

I learned an important lesson that day: Success in a large organization, whether it's a university or IBM, is generally based on appearance, not reality. It's understanding the system and then working within it that really counts, not bowling scores or body bags.

In the world of high-tech start-ups, there *is no system*, there are no hard and fast rules, and all that counts is the end product. The high-tech start-up bowling league would allow genetically engineered bowlers, superconducting bowling balls, tactical nuclear weapons—anything to help your score or hurt the other guy's. Anything goes, and that's what makes the start-up so much fun.

No wonder they turned the Stanford University bowling alley into a computer room.

What makes start-ups possible at all is the fact that there are lots of people who like to work in that kind of environment. And Americans seem more willing than other nationalities to accept the high probability of working for a company that fails. Maybe that's because to American engineers and programmers, the professional risk of being with a start-up is very low. The high demand for computer professionals means that if a start-up fails, its workers can always find other jobs. If they are any good at all, they can get a new job in two weeks. So that's the personal risk of joining a start-up: two weeks' pay.

Good thing, too, because most start-ups fail.

But they don't have to. Time for Bob Cringely's guide to starting your own high-tech company, getting rich, then getting out.

Conventional wisdom says that nine out of ten start-ups fail. My friend Joe Adler, who eschews conventional wisdom in favor of statistics, claims that the real numbers are even worse. He says that nineteen start-ups out of twenty fail. And since Joe has done both successful and unsuccessful start-ups and teaches a class about them at the Stanford Graduate School of Business, let's believe him.

If nineteen out of twenty start-ups fail, then it seems to me that the books on how to be successful in Silicon Valley are taking the wrong approach. My guide will let success take care of itself. Instead, I'll concentrate on the much harder job of how not to fail.

High-tech start-ups fail for only three reasons: stupidity, bad luck, and greed.

Starting a mainframe computer company in 1992 would be stupid. In general, starting a company to do any me-too product, any non-state-of-the-art product, or any product in a declining market would be stupid. My guess is that stupidity claims 25 percent of all start-ups, which would explain five of those nineteen failures. Fourteen to go.

No start-up I know of ever failed because of good luck, but bad luck takes as many companies as stupidity does—five out of twenty. Bad luck comes in the form of an unexpected recession that dries up funding. It often means the appearance of an unexpected rival, introducing a better product the month before yours is to be announced. And it even means getting loaded on the day your company goes public, driving your new Ferrari into a ditch, and getting killed, scotching the IPO. That's what happened to the founder of Eagle Computer, an early maker of PC clones.

Tip 1 for would-be entrepreneurs: *Avoid stupid and unlucky people. If you are stupid or have bad luck, don't start a high-tech business.*

That leaves us with greed, which I say causes at least half of all high-tech start-up failures. If we could eliminate greed entirely, ten out of twenty start-ups would succeed—ten times the current success rate.

Greed takes many forms but always afflicts company founders.

Say you want to start a company but can't think of a product to build. Just then a venture capitalist calls, looking for someone working on a spreadsheet program for the Acme X-14 computer, or maybe it's a graphics board for the X-14 or a floating-point chip. Anyway, the guy wants to invest $2 million, and all you have to do to get the money is tell him that's what you had in mind to work on all along.

Don't do it.

After the success of Compaq Computer, every venture capitalist in the world wanted to fund a PC clone company. After the success of Lotus Development, every venture capitalist in the world wanted to fund a PC software company. They threw tons of money at anyone who could claim anything like a track record. Those people took the money and generally failed because they were fulfilling some venture capitalist's dream, not their own.

We're talking pure greed here, on the part of both the venture capitalist and the entrepreneur. VCs love to do me-too products and have had a tendency to fund simultaneously twenty-six hard disk companies that all expect to have 8 percent of the market within two years. It doesn't work that way.

Tip 2 for would-be entrepreneurs: *Do a product that you want to do, not one that they want you to do.*

Or maybe you already know what your product will be, and one day a venture capitalist drops by, hears your idea, and offers you $2 million on the spot in exchange for a large percentage of the company.

Don't take it.

Start-up founders generally have only ideas, charisma, and equity to work with. Ideas and charisma are cheap, but equity is expensive. To make a start-up work, the founder has to divvy out parts of the business at just the right rate to keep everyone happy until the product is a success. Give away too much of your company too soon to a venture capitalist, to your co-workers, or even to yourself, and you risk running out of distributable shares before the product is done. And that probably means the product won't be done. Ever.

Tip 3 for would-be entrepreneurs: *Don't take venture funding too soon.*

If you are doing a software product, don't take venture money until you need it to introduce the product. If you are doing a hardware product, don't take venture money until you have used up all of your own money, your mother-in-law's money, and everything you can borrow.

Bootstrap. Rent; don't buy. Don't hire people to do things you can contract out because contractors don't require stock options. Don't hire marketers too soon because that will only dilute the equity pool available to the technical people who are finishing up the product. You don't want to alienate those guys.

In fact, you don't want to alienate anyone. As founder, your job is to keep everyone else happy by giving away your company. Give it away carefully, but give it away, because not doing so guarantees you will be the majority shareholder in a worthless enterprise. Don't be greedy.

As the founder, the man or woman with the grand plan, your function is to manage the distribution of your own holdings so that you end up with fewer shares but more wealth. The idea is to end up with a thinner slice of a thicker pie. When Bob Metcalfe started 3Com Corp. in June 1979, he owned 100 percent of nothing. When 3Com went public in March 1984, he owned 12 percent of a company with a fair market value of $80 million.

Tip 4 for would-be entrepreneurs: *Invite me to lunch. I'm a cheap date.*

There is an enormous difference between starting a company and running one. Thinking up great ideas, which requires mainly intelligence and knowledge, is much easier than building an organization, which also requires measures of tenacity, discipline, and understanding. Part of the reason that nineteen out of twenty high-tech start-ups end in failure must be the difficulty of making this critical transition from a bunch of guys in a rented office to a larger bunch of guys in a rented office with customers to serve. Customers? What are those?

Think of the growth of a company as a military operation, which isn't such a stretch, given that both enterprises involve strategy, tactics, supply lines, communication, alliances, and manpower.

Whether invading countries or markets, the first wave of troops to see battle are the commandos. Woz and Jobs were the commandos of the Apple II. Don Estridge and his twelve disciples were the commandos of the IBM PC. Dan Bricklin and Bob Frankston were the commandos of VisiCalc. Mitch Kapor and Jonathan Sachs were the commandos of Lotus 1-2-3. Commandos parachute behind enemy lines or quietly crawl ashore at night. A start-up's biggest advantage is speed, and speed is what commandos live for. They work hard, fast, and cheap, though often with a low level of professionalism, which is okay, too, because professionalism is expensive. Their job is to do lots of damage with surprise and teamwork, establishing a beachhead before the enemy is even aware that they exist. Ideally, they do this by building the prototype of a product that is so creative, so exactly correct for its purpose that by its very existence it leads to the destruction of other products. They make creativity a destructive act.

For many products, and even for entire families of products, the commandos are the only forces that are allowed to be creative. Only they get to push the state of the art, providing creative solutions to customer needs. They have contact with potential customers, view the development process as an adventure, and work on the total product. But what they build, while it may look like a product and work like a product, usually isn't a product because it still has bugs and major failings that are beneath the notice of commando types. Or maybe it works fine but can't be produced profitably without extensive redesign. Commandos are useless for this type of work. They get bored.

I remember watching a paratrooper being interviewed on televison in Panama after the U.S. invasion. "It's not great," he said. "We're still here."

Sometimes commandos are bored even before the prototype is complete, so it stalls. The choice then is to wait for the commandos to regain interest or to find a new squad of commandos.

When 3Com Corp. was developing the first circuit card that would allow personal computers to communicate over Ethernet computer networks, the lead commando was Ron Crane, a brilliant, if erratic, engineer. The very future of 3Com depended on his finishing the Ethernet card on time, since the company was rapidly going broke and additional venture funding was tied to successful completion of the card. No Ethernet card, no money; no money, no company. In the middle of this high-pressure assignment, Crane just stopped working on the Ethernet card, leaving it unfinished on his workbench, and compulsively turned to finding a way to measure the sound reflectivity of his office ceiling tiles. That's the way it is sometimes when commandos get bored. Nobody else was prepared to take over Crane's job, so all his co-workers at 3Com could think to do in this moment of crisis was to wait for the end of his research, hoping that it would go well.

The happy ending here is that Crane eventually established 3Com's ceiling tile acoustic reflectivity standard, regained his Ethernet bearings, and delivered the breakthrough product, allowing 3Com to achieve its destiny as a $400 million company.

It's easy to dismiss the commandos. After all, most of business and warfare is conventional. But without commandos, you'd never get on the beach at all.

Grouping offshore as the commandos do their work is the second wave of soldiers, the infantry. These are the people who hit the beach en masse and slog out the early victory, building on the start given them by the commandos. The second-wave troops take the prototype, test it, refine it, make it manufacturable, write the manuals, market it, and ideally produce a profit. Because there are so many more of these soldiers and their duties are so varied, they require an infrastructure of rules and procedures for getting things done—all the stuff that commandos hate. For just this reason, soldiers of the second wave, while they can work with the first wave, generally don't trust them, though the commandos don't even notice this fact, since by this time they are bored and already looking for the door.

The second wave is hardest to manage because they require a structure in which to work. While the commandos make success possible, it's the infantry that makes success happen. They know their niche and expend the vast amounts of resources it takes to maintain position, or to reposition a product if the commandos made too many mistakes. While the commandos come up with creative ways to hurt the enemy, giving the start-up its purpose and early direction, the infantry actually kill the enemy or drive it away, occupying the battlefield and establishing a successful market presence for the start-up and its product.

What happens then is that the commandos and the infantry

head off in the direction of Berlin or Baghdad, advancing into new territories, performing their same jobs again and again, though each time in a slightly different way. But there is still a need for a military presence in the territory they leave behind, which they have liberated. These third-wave troops hate change. They aren't troops at all but police. They want to fuel growth not by planning more invasions and landing on more beaches but by adding people and building economies and empires of scale. AT&T, IBM, and practically all other big, old, successful industrial companies are examples of third-wave enterprises. They can't even remember their first- and second-wave founders.

Engineers in these established companies work on just part of a product, view their work as a job rather than an adventure, and usually have no customer contact. They also have no expectation of getting rich, and for good reason, because as companies grow, and especially after they go public, stock becomes a less effective employee motivator. They get fewer shares at a higher price, with less appreciation potential. Of course, there is also less risk, and to third-wave troops, this safety makes the lower reward worthwhile.

It's in the transitions between these waves of troops that peril lies for computer start-ups. The company founder and charismatic leader of the invasion is usually a commando, which means that he or she thrills to the idea of parachuting in and slashing throats but can't imagine running a mature organization that deals with the problems of customers or even with the problems of its own growing base of employees. Mitch Kapor of Lotus Development was an example of a commando/nice guy who didn't like to fire people or make unpopular decisions, and so eventually tired of being a chief executive, leaving at the height of its success the company he founded.

First-wave types have trouble, too, accepting the drudgery

that comes with being the boss of a high-tech start-up. Richard Leeds worked at Advanced Micro Devices and then Microsoft before starting his own small software company near Seattle. One day a programmer came to report that the toilet was plugged in the men's room. "Tell the office manager," Leeds said. "It's her job to handle things like that."

"I can't tell her," said the programmer, shyly. "She's a woman."

Richard Leeds, CEO, fixed the toilet.

The best leaders are experienced second-wave types who know enough to gather together a group of commandos and keep them inspired for the short time they are actually needed. Leaders who rise from the second wave must have both charisma and the ability to work with odd people. Don Estridge, who was recruited by Bill Lowe to head the development of the IBM PC, was a good second-wave leader. He could relate effectively to both IBM's third-wave management and the first-wave engineers who were needed to bring the original PC to market in just a year.

Apple chairman John Sculley is a third-wave leader of a second-wave company, which explains the many problems he has had over the years finding a focus for himself and for Apple. Sculley has been faking it.

When the leader is a third-wave type, the start-up is hardly ever successful, which is part of the reason that the idea of *intrapreneurism*—a trendy term for starting new companies inside larger, older companies—usually doesn't work. The third-wave managers of the parent company trust only other third-wave managers to run the start-up, but such managers don't know how to attract or keep commandos, so the enterprise generally has little hope of succeeding. This trend also explains the trouble that old-line computer companies have had entering the personal computer business. These companies can see only the big picture—the way that PCs fit into their broad product line of large and small computers. They concentrate more on fitting PCs politely into the product line than on kicking ass in the market, which is the way successes are built.

A team from Unisys Corp. dropped by *InfoWorld* one day to brag about the company's high-end personal computers. The boxes were priced at around $30,000, not because they cost so much to build but because setting the price any lower might have hurt the bottom end of Unisys's own line of minicomputers. Six miles away, at Fry's Electronics, the legendary Silicon Valley retailer that sells a unique combination of computers, junk food, and personal toiletry items, a virtually identical PC costs less than $3,000. Who buys Unisys PCs? Nobody.

Then Bob Kavner came to town, head of AT&T's computer operation and the guy who invested $300 million of Ma Bell's money in Sun Microsystems and then led AT&T's hostile acquisition of NCR—yet *another* company that didn't know its PC from a hole in the ground. Eating a cup of yogurt, Kavner asked why we gave his machines such bad scores in our product reviews. We'd tested the machines alongside competitors' models and found that the Ma Bell units were poorly designed and badly built. They compared poorly, and we told him so. Kavner was amazed, both by the fact that his products were so bad and to learn that we ran scientific tests; he thought it was just an *InfoWorld* grudge against AT&T. Here's a third-wave guy who was concentrating so hard on what was happening inside his own organization that he wasn't even aware of how that organization fit into the real world or, for that matter, how the real world even worked. No wonder AT&T has done poorly as a personal computer company.

Here's something that happens to every successful start-up: things go terrifically for months or years, and then suddenly half the founders quit the company. This is pent-up turnover because people have stayed with the company longer than they might have normally.

Say normal turnover is 10 percent per year, which is low for most high-tech companies. If nobody leaves for the first five years because they would lose their stock options, it shouldn't be surprising to see a 50 percent departure rate when the company finally goes public or is acquired. For years, those people were *dying* to leave. And they are naturally replaced with a different kind of worker—third-wave workers who are attracted to what they view as a stable, successful company.

Reasons other than boredom and pent-up ambition cause early employees to leave successful young companies. As companies get bigger, they become more organized and process driven, which leads to more waste. Great individual contributors—first- and second-wave types—are very efficient. They hate waste and are good indicators of its presence. When the best people start to bail out, it's a sign that there is too much waste.

Companies go through other transformations as they grow. Sales volumes go up, and quality control problems go up too. Fighting software bugs and hardware glitches, getting the product right before it goes out the door, rather than having to fix it afterward sops up more and more money. And as volume grows, so does penetration into the population of unsophisticated users, who require more hand holding than did the more experienced first users of the product. Suddenly what was once an adventure is now just a job.

WordPerfect Corp., the top PC word processing software company, has a building in Orem, Utah, where 600 people sit at computer workstations with the sole purpose of answering technical questions phoned in by customers who are struggling to use the product. Typical WordPerfect customers make two such calls, averaging five minutes each, which means that when the founders of a five-person software start-up dream about selling 100,000 copies of their new application, they are also dreaming about (though usually they don't know it) spending at least 8.3 man-years on the telephone answering the same questions over and over and over again.

Of course, companies don't have to grow. Electric Pencil, the first word processing program for the Apple II, was the archetype for all word processing packages that followed, but its developer, a former Hollywood screenwriter, just got tired of all the support hassles and finally shut his company down. In 1978, Electric Pencil had 250,000 users. By 1981, it was forgotten.

Some companies limit their responsibilities by licensing their products to other companies and avoid dealing with end users entirely. Convergent Technologies started this way, building computers that were sold by other companies under other names. Convergent was acting as an original equipment manufacturer, or OEM. For reasons that would have made no sense at all to Miss Vermillion, my seventh-grade English teacher, building products that are sold by others is called "OEMing."

Microsoft started out OEMing its software, selling its languages and operating systems to hardware companies that would ship the Microsoft code out under a different name—Zenith DOS, for example—packed in with the computer.

In the software business, there is a strong trend toward small companies' handing over their products to be marketed by larger companies. The big motivator here is not just the elimination of support costs but also removing the need to hire salespeople, make marketing plans, and develop relations with distributors. It can be easier and even more profitable to have your astrology program published as Lotus Stargazer than as the Two Guys in a Garage Astrology Program.

Finally, there are software companies that elect to remain small but profitable by literally giving their products away—every mainframe software salesman's idea of hell. This PC-peculiar product category is called "shareware."

Shareware was invented by Andrew Fluegelman and Jim Button. Button had spent eighteen years working as an engineer for IBM in Seattle when he bought one of the first IBM PCs to use at home but then couldn't find a database program to run on it. In 1982, the most popular database program was dBase II, which ran under CP/M, but there were no databases yet for the IBM PC.

Technical types who start software companies are either computer junkies who want to be the next Bill Gates (most are this type) or who need a program that isn't available so they write it themselves. Jim Button was from the latter group. His simple database program—PC File—became a hit with friends and co-workers in the Seattle area.

Friends asked for copies of the program, then those friends made copies for *their* friends, and soon there were dozens, maybe hundreds, of copies of PC File floating around the Pacific Northwest. This was fine except that these many nameless users sometimes had trouble making the program do what they wanted, so they tended to call Jim Button at home in the evenings with their questions, which came to require a lot of effort.

Button wanted to cut down his product support load, so he came up with the idea to put a simple message on the first screen of the program, telling users that they could get updates and improvements to PC File by sending $10 to Jim Button. Shareware was born.

The beauty of shareware was that there was no packaging, no printing, no marketing, no sales effort of any kind. The manual was included as a text file on the program disk; if users wanted it printed, they printed it themselves. Shareware was pure thought, just as if Jim Button dropped by the customer's house to give a demonstration of his programming prowess, only the real Jim Button was home in bed. Rather than go to a store or order by mail, users passed the programs around or got them over the telephone from computer bulletin boards. They tried it, and, if they liked it, maybe they sent Jim Button his $10 (later more). Having got the $10, Button sent on the next improved release of the product, which cost him maybe $2 for the floppy disk, envelope, and postage. He answered any questions from registered users and hoped to have the same customers paying him $10 every six to nine months as each new version of the product was shipped out with a few new features.

Button invented shareware during a time of hostile relations between sellers and users of software. The issue was copy protection. Software vendors didn't want ten bootlegged copies of each program to be floating around the country for each legal copy, and so they devised all sorts of technical tricks to make it harder for users to make copies of programs—tricks that alienated users in the process. Warning labels on the copy-protected diskettes said, generally, "Copy this product and we'll sue you, we'll take your youngest child, and end your productive life, dear customer." But Jim Button actually encouraged users to make copies of PC File for their friends. And if the friends didn't like the program or didn't feel that they needed their questions answered, they could easily get away without sending Button his $10.

He started a company he called Buttonware, operating out of his basement on evenings and weekends, funded by those $10 checks. Button drafted his wife and son to help with duplicating and shipping floppy disks while he worked on improving the program.

Button's fantasy, when he started asking for the $10 fee, was that the money would cover his time and eventually pay for a new computer. It went much further than that. Buttonware grew so fast that the Button family soon had no spare time at all, and Jim Button had to make a decision between giving up the home business or his career writing mainframe software for IBM. The decision came down to a simple economic analysis, made in the summer of 1984. Button looked at his 1984 salary for working at IBM, which was $50,000, and compared it to his earnings from Buttonware in the previous year, which were $490,000. Bye-bye Big Blue.

The price of PC File went to $25 when Andrew Fluegelman suggested they coordinate pricing on this new product category, which they were then calling Freeware. Fluegelman's product was a data communication program called PC-Talk that allowed PCs to emulate computer terminals and link to mainframes over telephone lines. The former corporate lawyer and editor of the *Whole Earth Catalog* wrote PC-Talk when he found that the communication program supplied with PC-DOS would not allow him to print from the screen while he was connected to an online information service.

Soon there were hundreds of other shareware programs. Bob Wallace, another Seattle programmer who was one of the first half-dozen Microsoft employees back in the Albuquerque days, wrote PC Write, the first shareware word processing package. Procomm was another communication package, this time coming from a company called Datastrom in Columbia, Missouri. Each of these hobby products eventually turned into full-time businesses with annual sales in the $2 million to $3 million range.

Price points were gradually raised, with each entrepreneur wondering when users would find it too expensive to register. Jim Button saw growth flatten when he reached $89, and Bob Wallace made the same discovery. Each man had to decide, then, whether just to control costs and milk profits from their products or to start marketing them finally. Both made the decision to grow, which meant spending money to create a more professional-looking product, advertising for the first time, and finding outlets other than shareware.

The trend in shareware companies is always the same. In the first few years, they grow to meet their destinies. If the product is good, it eventually fills the shareware channel, reaching all likely customers, at which point the companies look for growth through selling upgrades. But even upgrades eventually fade as users reach the point where their needs are served and adding two more esoteric features is not enough to compel them to pay for a $35 upgrade. At that point, while shareware sales are flat, the product has actually reached only 20 to 30 percent of the total software market, with 70 to 80 percent of potential users never having seen or heard of the program. Then it's time to try to find new channels of distribution. Jim Button tried retail stores, while Bob Wallace tried direct sales to large corporations, and each was successful. Datastorm made deals with hardware manufacturers to ship copies of Procomm bundled with the modems required for computer data communication.

Or maybe it's not time to grow. That's the other choice that many shareware publishers make—the types who want to stay small, working by themselves, and just make a good living mining some tiny software niche in the vast MS-DOS marketplace. Astrology software, anyone?

ECONOMICS OF SCALE

We're at the ballpark, now, and while you and I are taking a second bite from our chilidogs, this is what's happening in the outfield, according to Rick Miller, a former Gold Glove center fielder for the Bosox and the Angels. When the pitcher's winding up, and we figure the center fielder's just stooped over out there, waiting for the photon torpedoes to load and thinking about T-bills or jock itch endorsements, he's really watching the pitcher and getting ready to catch the ball that has yet to be thrown. Exceptional center fielders use three main factors in judging where the ball will land: what kind of pitch is thrown where in the hitter's zone, the first six inches of the batter's swing, and the sound of the ball coming off the bat.

So Miller watches, then listens, then runs. Except for the most routine of hits, he never looks up to see the ball until he gets to where it is going to land; he just moves to where it *should* land. This technique works well except at indoor ballparks like the Seattle Kingdome. The acoustics in the Kingdome are such that Miller has to watch the ball for the half-second after it leaves the bat, just like the rest of us would do, and it costs about 20 percent of his range.

I will never be a Rick Miller. Bob Cringely, the guy who says it shouldn't take six years to learn to be a blacksmith, wasn't talking about what it would take to be the world's best blacksmith. I could start today taking Rick Miller lessons from Rick Miller, and in six years or even sixty years could never duplicate his skills. It's a bummer, I know, but it's just too late for me to make the major leagues. Or even the Little League.

Back in elementary school, when all the other boys were shagging flies and grounders until sundown, I must have been doing something else. For some reason—I don't remember what I was doing instead—I never played baseball as a kid. And because I never played baseball, I'll always be in the stands eating chilidogs and never be in center field being Rick Miller.

There's only one way to be a Rick Miller, and that's to start training for the job when you are 8 years old. Ten years and 200,000 pop flies later, you are ready for the minor leagues. Three years after that, it's time for the majors—the show. There are no short-cuts. A robot, a first-string goalie from the New York Rangers, or a genetically engineered boy from Brazil could not come into the game as an adult and hope to be a factor.

Even if Rick Miller himself was doing the teaching, it wouldn't work. He'd say, "Hear the way the bat sounds? Quick, run to the right! Hear that one? Run to the left! This one's going long! Back! Back! Back!"

But they'd all sound the same to you and me. We'd have to hear the sounds and learn to make the associations ourselves over time. We'd need those 200,000 fly balls and the ten years it would take to catch them all.

There is no substitute for experience. And except for certain moves that I surprised myself with one evening years ago in the back seat of a DeSoto, there are no skills or knowledge that just spontaneously appear at a certain preprogrammed point in life.

My mother is unaware of this latter point. She bought me

white 100 percent cotton J.C. Penney briefs for the first eighteen years of my life and then was surprised during a recent visit to learn that I hadn't spontaneously switched to boxer shorts like my dad's. She just assumed that there was some boxer short gene that lay dormant until making itself known to men after high school. There isn't. I still wear white 100 percent cotton J.C. Penney briefs, Mom. I probably always will.

And now we're back in the personal computer business, where there is also no substitute for experience, where good CEOs do not automatically generate from good programmers or engineers, and where everything, including growth, comes at a cost.

≈≈

For computer companies, the cost of growth is usually innocence. Many company founders, who have no trouble managing twenty-five highly motivated techies, fail miserably when their work force has grown to 500 and includes all types of workers. And why shouldn't they fail? They aren't trained as managers. They haven't been working their way up the management ladder in a big company like IBM. More likely, they are 30 years old and suddenly responsible for $30 million in sales, 500 families, and a customer base that keeps asking for service and support. Sometimes the leader, who never really imagined getting stuck in this particular rut, is up to the job and learns how to cope. And sometimes he or she is *not* up to the job and either destroys the company or is replaced with another plague—professional management.

There comes a day when the founders start to disappear, and the suits appear, with their M.B.A.s and their ideas about price points, market penetration, and strategic positioning. And because these new people don't usually understand the inner workings of the computer or the software that is the stuff actually made by the company they now work for, the nerds tend to ignore them, thinking that the suits are only a phase the company is going through on its way to regaining balance and remembering that engineers are the appropriate center of the organization.

The nerds look on their nontechnical co-workers—the marketing and financial types—as a necessary evil. They have to be kept around in order to make money, though the nerds are damned if they understand what these suits actually do. The techies are like teenagers who sat in the audience of the "Ed Sullivan Show," watching the Beatles or the Rolling Stones; the kids couldn't identify with Ed, but they knew he made the show possible, and so they gave him polite applause.

But the coming of the suits is more than a phase; it's what makes these companies bigger, sometimes it's what kills them on the way to being bigger, but either way it changes the character of each company and its leaders forever.

The great danger that comes with growth is losing the proper balance between technology and business. At the best companies, suits and nerds alike see themselves as part of a greater "us." That's the way it was at Lotus before the departure of Mitch Kapor. Kapor could use his TM training and his Woodstock manner to communicate with all types. As Lotus grew and some products were less successful than expected, Kapor found that the messages he was sending to his workers were increasingly dark and unpleasant. Why be worth $100 million and still have the job of giving people bad news? So Mitch Kapor gave up that job to Jim Manzi, who was 34 at the time, a feisty little guy from Yonkers who was perfectly willing to wear the black hat that came with power. But Manzi as CEO lacked understanding of the technology he was selling and the people he was selling it with.

Manzi was Lotus's first marketing vice-president, and he was the one who came up with the idea of marketing 1-2-3 directly to

corporations, advertising it in business and general interest publications that corporate leaders, rather than computer types,
might read. The plan worked brilliantly, and 1-2-3's success was a
phenomenon, selling $1 million worth in its first week on the
market. But for all his smarts, Manzi was also a suit in the strong-
est possible sense. He *sold* 1-2-3 but didn't use it. He boasted about
his lack of technical knowledge as though it was a virtue not to
understand the workings of his company's major product. His
position was that he had people to understand that stuff for him.
Being able to sell software so brilliantly while lacking a technical
understanding of the product was supposed to make him look all
the smarter, a look Manzi wanted very much to cultivate.

While he was totally reliant on people to explain the lay of
the computer landscape, Manzi didn't know any more about
how to use people than he did 1-2-3. Five development heads
came and left Lotus in four years, and each of these technical
leads consistently went from making Manzi "ecstatic" with their
progress to being "dickheads." Programming went from being
down the hall to "in the lab," which could just as well have been
in another country, since Manzi had no idea what was going on
there, and his technical people felt no particular need to share
their work with him either. At least three major products that
would come to have bottom-line importance for Lotus were
developed without Manzi's even knowing they existed because
of his isolation from the troops.

When all of Manzi's emphasis was on 1-2-3 version 3.0, the
advanced spreadsheet that was delayed again and again and
would not be born, a couple of programmers working on their
own came up with 1-2-3 version 2.2, a significant improvement
over the version then shipping. By the time Manzi even knew
about 2.2, its authors had quit the company in disgust, leaving
behind their code, which eventually made millions for Lotus
when it was finally discovered and promoted.

"May I join you?" Manzi once asked a group of Lotus employees in the company cafeteria, "or do you hate me like everyone else?"

Poor Jimmy.

"Manzi is a bad sociopath—one that is incapable of using friends," claimed Marv Goldschmitt, who ran Lotus's international operations until 1985. "A good sociopath manipulates and therefore needs to have people around. Manzi, as a bad sociopath, sees people inside Lotus as enemies. He could have kept a lot of good people who left the company—and he should have but saw them as dangerous."

This attitude extended even to strategic partners. When Compaq Computer used some of his remarks in a promotional video without his permission, Manzi tore apart his own Compaq computer, stuffed it in a box, and shipped the parts directly to Rod Canion, Compaq's CEO, with a note saying he didn't want the thing on his desk anymore.

With 1-2-3 the largest-selling MS-DOS application, it would have been logical for Manzi to have had a good relationship with Microsoft's Bill Gates. Nope. Having barely escaped being acquired by Microsoft back in 1984, Manzi had no good feelings for Gates. He specifically tried to keep Lotus from developing a spreadsheet to work under Microsoft's Windows graphical environment, for example, because he did not want to do anything to assist Gates. But trying to stop a product from happening and actually doing so were different things. Down in the lab, even as Manzi railed against Windows, was Amstel, a low-end Windows spreadsheet developed at Lotus without Manzi's ever being aware of it. Amstel eventually turned into 1-2-3/Windows, an important Lotus product.

Manzi saw himself in competition with Gates. Each man wanted to be head of the biggest PC software company. Each wanted to be infinitely rich (though only Gates was). They even

competed as car collectors. Gates and Paul Allen dropped
$400,000 each into a pair of aluminum-bodied Porsche 959
sports cars, so Manzi also ordered one, even though the cars were
never intended to be sold in the United States. Allen and Gates
took delivery of serial numbers 197 and 198, and Manzi would
have got number 201 except that Porsche decided to stop production at 200. Beaten again by Bill Gates.

Alienated by choice from the rest of his company, Manzi
churned the organization with regular reorganizations, claiming he was fostering innovation but knowing that he was also
making it harder for rivals to gain power. Taciturn, feeling so
unlovable that he could not trust anyone, Manzi created development groups of up to 200 people, knowing they would be
hard to organize against him. Such large groups also guaranteed that new versions of 1-2-3 would be delayed, sometimes for
years, as communication problems overwhelmed the large
numbers of programmers.

The bad news about Lotus was slow in coming because the
installed base of several million users kept cash flowing long
after innovation was stifled. In 1987, right in the middle of this
bleak period, Manzi earned $26 million in salary, bonuses, and
stock options. But the truth always comes out, and in the case of
Lotus, even Manzi eventually had to take a chance and trust
someone, in this case Frank King, an old-line manager from IBM
who definitely did understand the technology.

Frank King had been the inventor of SQL, an innovative
database language that somehow appeared from the catacombs
of IBM. Like nearly every other clever product from IBM, SQL
had been developed in secret. King and his group developed
SQL in a closet, lied about it, then finally showed it to the big-
shots who were too impressed to turn the product down. Frank
King knows how to get things done.

It was King who set up five offices at Lotus, one in every development group, and spent a day per week in each. It was King who discovered the hidden products that had been there all along and who got the long-delayed, though still flawed, Lotus 1-2-3 3.0 unstuck. It was Mitch Kapor and Jim Manzi who made Lotus and Frank King who saved it.

In a company with a strong founder, power goes to those who sway the founder. In most companies, this eventually means a rise of articulate marketers and a loss of status for developers. That's what happened at Aldus, inventors of desktop publishing and PageMaker, which turned out to be the compelling application for Apple's Macintosh computer.

Aldus was founded by a group of six men who had split away from Atex, a maker of minicomputer-based publishing systems for magazines and newspapers. Atex had an operation in Redmond, Washington, devoted to integrating personal computers as workstations on its systems. When Massachusetts-based Atex decided to close the Redmond operation, Paul Brainerd, who managed the Washington operation, recruited five engineers to start a new company. They set out to invent what came to be called desktop publishing. Brainerd contributed his time and $100,000 to the venture, while the five engineers agreed to work for half what they had been paid at Atex.

Aldus was originally pitched as a partnership, but, typically, the engineers didn't pay attention to those organization things. That changed one day when they all met at the courthouse to sign incorporation papers and the others discovered that Brainerd was getting 1 million shares of stock while each of the engineers was getting only 27,000 shares. Brainerd was taking

95 percent of the stock in the company giving the others 1 percent each. The techies balked, refused to sign, and eventually
got their holdings doubled. For his $100,000, Brainerd bought
90 percent of Aldus.

Paul Brainerd was into getting his own way.

"It's common for founders of these companies to be abusive,"
said Jeremy Jaech, one of the five original Aldus engineers. "Certainly Brainerd, Jobs, and Gates are that way. I looked up to Paul
as a father figure, and so did most of the other founders and early
staff. I was 29 when we started, and most of the others were even
younger. We came to see Paul as the demanding father who could
never be pleased. It was like a family situation where, years later,
you wonder how you let yourself get so jerked around over what,
in retrospect, seems to be so unimportant. 'Why did I care so
much [about what he thought]?' I keep asking myself."

Brainerd's money lasted six months, long enough to build a
prototype of the application and to write a business plan. The
first prototype was finished in three months; then Brainerd went
on the road, making his pitch to forty-nine venture capitalists
before finding his one and only taker. The plan had been to raise
$1 million, but only $846,000 was available. It was just enough.

It wasn't clear how venture capitalists could assign a value to
software companies, so they tended to shy away from software,
thinking that hardware was somehow more certain. The VCs were
always worried that someone else writing software in another
garage would do the same thing, either a little bit quicker or a
little bit better. The money people were so uninterested that Brainerd found that most of the VCs, in fact, hadn't even read the
Aldus business plan.

You need a big partner to start a new product niche in the
personal computer business. For Aldus, the partner was Apple,
which needed applications to help it sell its expensive LaserWriter
printer. Apple's dealers had been burned by the failure of the Lisa,

the HP Laserjet printer was out on the market already and much cheaper, and no software was available that used LaserWriter's PostScript language. The situation didn't look good. Apple was worried that the LaserWriter would bomb. Apple *needed* Aldus. Three LaserWriter prototypes were given to software developers in September 1984. One went to Lotus, one to Microsoft, and one to Aldus, so Apple had a clear sense of the potential importance of PageMaker, the first program specifically for positioning text and graphics on a PostScript printed page.

Aldus's original strategy was to show dealers that PageMaker would sell hardware. They kept the number of dealers small to avoid price cutting. The early users were mainly small business-people. Compared to going outside to professional typesetters to prepare their company newsletters and forms, PageMaker saved them time and money and gave them control of the process. It was this last part that actually drove the sale. Traditional typesetting businesses didn't pay much attention to customers, so small businesspeople were alienated. With Pagemaker and a Laser-Writer, they no longer needed the typesetters.

Aldus surprised the computer world by taking what everyone thought was a vertical application—an application of interest only to a specialized group like professional typesetters—and showed that it was really a horizontal application—an application of interest to nearly every business. Companies didn't produce as many newsletters as spreadsheets, but nearly all produced at least one or two newsletters, and that was enough to make the Macintosh a success. There was nothing like PageMaker and the Laser-Writer in the world of MS-DOS computing.

The first release of Pagemaker was filled with bugs, but microcomputer users are patient, especially with groundbreaking applications. There was talk inside the company of holding PageMaker back for one more revision, but the company was out of money and that would have meant going out of business. Like most other products from software start-ups, PageMaker was shipped when it had to be, not when it was done. Three months later, a second release fixed most of the bigger problems.

By the late 1980s, Aldus was a success, and Paul Brainerd was a very wealthy man. But Brainerd was trapped too. When Aldus was started, the stated plan was to work like hell for five years and then sell out for a lot of money. That's the dream of every start-up, but it's a dream that doesn't hold up well in the face of reality. Brainerd had discussions with Bill Gates about selling out to Microsoft, but those talks failed and Aldus had no choice but to go public and at least pretend to grow up.

Companies used to go public to raise capital. They needed money to build a new steel mill or to lay a string of railroad track from here to Chicago, and rather than borrow the money to pay for such expansion, they sold company shares to the investing public. That's not why computer companies go public.

Computer companies generally don't need any money when they go public. Apple Computer was sitting on more than $100 million in cash when it went public in 1979. Microsoft had even more cash than that stashed away when it went public in 1986. These numbers aren't unusual in the hardware and software businesses, which have always been terrific cash generators.

It's not unusual at all for a software company with $50 million in sales to be sitting on $30 million to $40 million in cash. Intel these days has about $8 billion in sales and $2 billion in cash. Microsoft has $2.8 billion in sales and more than $900 million in cash. Apple, with $8 billion in sales, is sitting on a bigger pile of cash than the company will even admit to. At the same time it is laying off workers in the United States and moaning about flat or falling earnings, Apple admits to having $1 billion in cash in the United States, and has at least another billion stashed overseas, with no way to bring it into the United States without paying a lot of taxes. None of these companies has a dime of long-term debt.

This habit of sitting on a big pile of money originated at Hewlett-Packard in the 1940s. David Packard figured that careful management of inventories and cash flow could generate lots of money over time. Hanging on to that money meant that the next emergency or major expansion could be financed entirely from internal funds. Now every company in Silicon Valley manages its finances the H-P way.

What's ironic about all these bags of money lying around the corporate treasuries of Silicon Valley is that although the loot provides insurance for hard times ahead, it actually drags down company earnings. "Sure, I've got $600–700 million available, but who needs it?" asked Frank Gaudette, Microsoft's chief financial officer. "I've got to find places to put the money, and then what do I make—12–15 percent, maybe? Better I should churn the money right back into the company, where we average 40 or 50 percent return on invested capital. We're losing money on all that cash."

But not even Microsoft can grow fast enough to absorb all that money, so the excess is often used to buy back company stock. "It increases the value of the outstanding shares, which is like an untaxed dividend for our shareholders," Gaudette said.

While computer companies are aggressive about managing their cash flow, they are usually very conservative about their tax accounting. Most personal computer software companies, for example, don't depreciate the value of their software; they pretend it has no value at all. IBM carries more than $2 billion on its books as the depreciable value of its software. Microsoft carries no value on its books for MS-DOS or any of its other products. If Microsoft managed its accounting the way IBM does, its earnings would be twice what they are today with no other changes required. *That's* why Wall Street loves Microsoft stock.

So computer companies don't go public to raise money; they go public to make real the wealth of their founders. Stock options are worthless unless the stock is publicly traded. And only when the stock is traded can founders convert some of their holdings in Acme Software or Acme Computer Hardware into the more dull but durable form of T-bills and real estate—wealth that has meaning, that makes it worthwhile for cousins and grandnephews to fight over after the entrepreneur is dead.

Bill Gates never wanted to take Microsoft public, but all those kids who'd worked their asses off for their 10,000 shares of founders' stock wanted to cash out. These early Microsoft employees—the ones walking around wearing FYIFV lapel buttons, which stand for *Fuck You, I'm Fully Vested*—were millionaires on paper but still unable to qualify for mortgages. They started selling their Microsoft shares privately, gaining the attention of the SEC, which began pushing the company toward an initial public offering. Gates eventually had no choice but to take Microsoft public, making himself a billionaire in the process.

Companies that don't grant stock options to employees have no trouble staying private, of course. That's what happened at WordPerfect Corp., the leading maker of PC word processing software. Started in Utah by a Brigham Young University computer science professor in partnership with the director of the BYU marching band, WordPerfect now has more than $300 million in annual sales yet only three stockholders. The company also has more than $100 million in cash.

Paul Brainerd was one of those founders who wanted to stabilize his fortune, giving his kids something to fight over. Overnight, Brainerd became very rich by making a public company of Aldus Corp. But Brainerd's secure fortune, like that of every other entrepreneur turned CEO of a public company, came at a personal cost.

Start-ups are built on the idea of working hard for five years and then selling out, but public companies are supposed to last forever. CEOs of public companies stand before analysts and shareholders, promising ever higher earnings from now until the end of time. Like other entrepreneurs-turned-corporate honcho, Brainerd is rich, but he's also trapped at Aldus, by both money and ego. His enormous holdings mean that it would take too long to sell all that stock unless he sells the whole company to a larger firm. And there is an emotional cost, too, since he believes that he can't do it again. This is his chance to be a big shot. Brainerd has a large ego. He needs power, and if he left Aldus, what would he do?

There are two kinds of software companies; one develops new concepts and pioneers new product areas, and the other works at continuing the evolution of an existing product. These two types of companies, and the people they need to do their jobs well, are very different. Aldus used to be the first type, but today it is very much the second type of company, and the people of Aldus have had to change to fit. Their primary job is to keep improving PageMaker. Public companies with successful products put their money into guaranteed winners, which means upgrades to the core product and add-on programs for it. At Aldus today, all the other products are viewed as supplements to PageMaker, which must be protected. PageMaker is the cash cow.

Aldus programmers concentrate on new versions of PageMaker, while most other applications sold under the Aldus name are actually bought from outside developers. Freehand, a drawing package, came from a company in Texas called Altsys, which gets a 15 percent royalty on sales. Persuasion, a package for automating business presentations, is another Aldus product gotten from outside, this time with a 12 percent royalty but a bigger down payment. Although it pays 15 percent royalties for products developed outside, Aldus, like most other established software companies, budgets only 6 or 7 percent of sales for internal development projects. This is frustrating for the programmers inside because they are responsible for the vast majority of sales yet are budgeted at a rate only half that of acquired products. Aldus expects more of them yet gives them fewer resources.

Successful software companies like Aldus quickly become risk averse. They buy outside products for lots of money with the idea that they are buying only good, already completed products that are more likely to succeed. Internal development of new products suffers because of the continual need to revise the cash cow and because the company is afraid of spending too much money developing duds.

For an example of such risk aversion, consider Aldus's abortive entry into the word processing software market. Although PageMaker was a desktop publishing program, it originally offered no facility for inputting text. Instead, it read text files from other word processing packages. When Aldus was working on PC PageMaker, which would run under Microsoft Windows on MS-DOS PCs, it seemed logical to add text input, and even to develop Aldus's own word processing package for Windows. Code-named Flintstone, the Aldus word processor would have had a chance to dominate the young market for Windows word processors.

By early 1988, a prototype of Flintstone was running, though it was still a year from being ready to ship. That's when Bill Gates gave Paul Brainerd a demonstration of Word for Windows—Microsoft's word processor that would compete with Flintstone. Gates told Brainerd that Word for Windows would ship in six to nine months, beating Flintstone to market. Afraid of going head to head against Microsoft, Brainerd canceled Flintstone. Word for Windows finally hit the market two years later.

While Lotus was a technology company with good marketing that became a marketing company with okay technology, some computer and software companies have always been marketing organizations, dependent on technology from outside. Even these firms can run aground from problems of growth and the transition of power.

Look at Ashton-Tate. George Tate's three-person firm contracted in 1980 to market Wayne Ratliff's database program called Vulcan. Vulcan was a subset of a public domain database called JPLDIS that Ratliff, an engineer at Martin-Marrietta Corp., had used on mainframe computers running at the Jet Propulsion Laboratory in Pasadena. Some have claimed that Ratliff wrote JPLDIS, but the truth is that he only wrote Vulcan, which had a subset of JPLDIS features combined with a full-screen interface, allowing users to seek and sort data by filling out an on-screen form rather than typing a list of cryptic commands.

Ratliff tried selling Vulcan himself, but the load of running a one-man operation while still working at Martin-Marrietta during the day was wearing. Rather than quit his day job, Ratliff pulled Vulcan from the market, later selling marketing rights to George Tate. The product was renamed dBase II and became the most successful microcomputer database program of its time. Ratliff, who had hoped to earn a total of $100,000 from his relationship with Tate, made millions.

Ratliff worked for Martin-Marrietta until 1982 while continuing to develop dBase II in his spare time, as required by his contract with Ashton-Tate. There was no program development at all done at Ashton-Tate's headquarters in Torrance, which was strictly a marketing and finance operation. By 1983, when introduction of the IBM PC-XT with its hard disk drive made clear how big a success dBase II was going to be in the PC-DOS market, Tate bought rights to the program outright and installed Ratliff in Torrance as head of development for dBase III.

It was at this time, when dBase III was as successful in the database market as Lotus 1-2-3 was among spreadsheets, that

George Tate snorted one line of cocaine too many and died of a heart attack at his desk. Suddenly Ashton-Tate had a new CEO, Ed Esber, who had been hired away from Dan Fylstra's VisiCorp to be marketing vice-president only a few weeks before. Esber, who was 32, was a marketer, not a technologist, and except for the vacuum created by Tate's sudden death probably would not have been considered for the jobs of president, chairman, and CEO that fell to him.

In his new position, Esber made the mistake of tipping the balance of power too much in the direction of marketing, then toward finance, and all at a major cost in lost time and bad technology. Marketing figured out what the next program was supposed to do; detailed specifications were written and then distributed to a large number of programmers, who were expected to write modules of code that would work together. Only they didn't work together, at least not well, in part because the marketers didn't have a clear concept of what was possible and what wasn't when the specs were written. These were marketers acting as metaprogrammers and not knowing what the hell they were doing.

Ashton-Tate began to have the same problems bringing out its next version of dBase—dBase IV—that Lotus was having with 1-2-3 version 3.0. The company bought outside products like Framework, an integrated package that competed with 1-2-3, and MultiMate, a word processor, but even these were allowed to bog down in the bureaucracy that resulted from an organization whose leaders didn't know what they were doing.

"Esber thought management of a development group meant going over the phone bills and accusing us of making too many long-distance calls," said Robert Carr, who wrote Framework and was Ashton-Tate's chief scientist in those days.

When dBase IV finally shipped, it was nearly two years late. Worse, it didn't work well at all. The product was seriously flawed and the programmers knew it. Still, the product was shipped because the finance-oriented company was worried about declining cash flow. They shipped dBase IV only to help sales and earnings. But bad software is its own reward; the resulting firestorm of customer complaints nearly drove the company out of business.

Ratliff left, and competitors like Nantucket Software and Fox Software created dBase-like programs and dBase add-ons that outperformed the original. Despite having 2.3 million dBase users and over $100 million in the bank, Esber was forced out during the spring of 1990 when Ashton-Tate posted a $41 million loss.

The week after he was pushed from power, Ed Esber had his first-ever dBase programming lesson.

∽∽

The suits first appeared at Microsoft in 1980, right around the time of the IBM deal. Prior to that time, Microsoft was strictly a maker of OEM software sold to computer companies and maybe to the occasional large corporation. Those corporate deals were simple and often clumsily done. In 1979, for example, Microsoft gave Boeing Commercial Airplane Co. the right to buy any Microsoft product for $50 per copy, *until the end of time*. Today most Microsoft applications sell in the $300 to $500 range, ten years from now they may cost thousands each, but Boeing still would be paying just $50.

When Microsoft realized its mistake, a blonde suit in her twenties named Jennifer Seman was sent alone to do battle with Boeing's lawyers. First she dropped the Boeing contract off with Microsoft's chief counsel for a legal analysis; when she came back a few days later to talk about the contract, it was on the floor, underneath one leg of the lawyer's chair, still unread.

That was the way they did things when Microsoft was still small, when what people meant when they said "Microsoft" was a group of kids wearing jeans and T-shirts and working in a cheap office near the freeway in Bellevue. The programmers weren't just the center of the company in those days, they *were* the company. There was no infrastructure at all, no management systems, no procedures.

Microsoft wasn't very professional back then. A typical Microsoft scene was Gordon Letwin, a top programmer, invading the office of Vern Raburn, head of sales, to measure it and find that Raburn's office was, as suspected, *three inches larger* than Letwin's. Microsoft was a company being run like a fraternity, and, as such, it made perfect sense when one hacker's expense account included the purchase of a pool table. Boys need toys.

But Bill Gates knew that to achieve his goals, Microsoft would have to become a much larger company, with attendant big company systems. He didn't know how to go about creating those systems, so he hired a president, Robert Towne, from an electronics company in Oregon called Tektronix, and a marketing communications whiz, Roland Hansen, who had been instrumental in the success of Neutrogena soap.

Towne lasted just over a year. The programmers quickly identified him as a dweeb, and ignored him. Gates continually countermanded his orders.

Hansen's was a different story. He dealt in the black magic of image and quickly realized that the franchise at Microsoft was Bill Gates. Hansen's main job would be to make Gates into an industry figure and then a national figure if Microsoft was to become the company its founder imagined it would be. The alternative to Gates was Paul Allen, but the co-founder was too painfully shy to handle the pressure of being in the public spotlight, while Gates looked forward to such encounters. Paul Allen's idea of a public persona is sitting with his mother in front-row seats for home games of his favorite possession, the Portland Trailblazers of the NBA.

Even with Gates, Hansen's work was cut out for him. It would be a challenge to promote a nerd with few social skills, who was only marginally controllable in public situations and sometimes went weeks without bathing. Maybe Neutrogena soap was a fitting precedent.

To his credit, by 1983 Hansen managed to get Gates's face on the cover of *Time* magazine, though Gates was irked that Steve Jobs of Apple had made the cover before he did.

Massaging Bill's image did nothing for organizing the company, so Gates went looking for another president after Towne's departure. By this time, Paul Allen had left the company, suffering from Hodgkin's disease, and Gates was in total control, which meant, in short, that the company was in real trouble. Fortunately, Gates seemed to know the peril he was in and hired Tandy Corporation's Jon Shirley to be the new president of Microsoft. Shirley was not a dweeb.

Gates had been Microsoft's Tandy account manager when Shirley was head of the Radio Shack computer merchandising operation. Although Shirley had made mistakes at Tandy, notably deciding against 100 percent IBM compatibility for its PC line, that didn't matter to Gates, who wasn't hiring Shirley for his technical judgment. Technology was Gates's job. He was hiring Shirley because he had successfully led the expansion of Tandy's Radio Shack stores across Europe. Shirley, who joined Radio Shack when he was a teenager, had literally watched Charles Tandy build the chain from the ground up to 7,000 stores worldwide. Shirley was to management what Rick Miller was to center field. Growing up at Radio Shack meant that Shirley knew about organization, leadership, and planning—things that Bill Gates knew nothing about.

Shirley's job was to build a business structure for Microsoft that both paralleled and supported the product development organization being built by Gates based on Simonyi's model. The trick was to create the systems that would allow the company to grow without diverting it from its focus on software development; Microsoft would ideally become a software development company that also did marketing, sales, support, and service rather than a marketing, sales, support, and service company that also developed software. This idea of nurturing the original purpose of the company while expanding the business organization is something that most software and hardware companies lose sight of as they grow. They managed it at Microsoft by having the programmers continue to report to Bill Gates while everyone on the business side reported to Shirley.

This was 1983. Microsoft was the second largest software company in the PC industry, was incredibly profitable, was growing at a rate of 100 percent per year, and had no debt. Microsoft was also a mess. There was no chief financial officer. The only company-wide computer system was electronic mail. Accounting systems were erratic. The manufacturing building was the only warehouse. The company was focused almost entirely on doing whatever the programmers wanted to do rather than what their customers were willing to pay for them to do.

One example of Microsoft's getting ahead of its customers' needs was the Microsoft mouse, which Gates had introduced not knowing who, if anyone, would buy it. At first nobody bought mice, and when Shirley started at Microsoft, he found a seven-year supply of electronic rodents on hand.

Then there was Flight Simulator, the only computer game published by Microsoft. There was no business plan that included a role for computer games in Microsoft's future. Bill Gates just liked to play Flight Simulator, so Microsoft published it.

In one day, Shirley hired a chief financial officer, a vice-president of manufacturing, a vice-president of human resources, a head of management information systems, and a head of investor relations. They were all the same person, Frank Gaudette, a wisecracking New Yorker hired away from Frito-Lay, who at 48 became Microsoft's oldest employee. Six years later, Gaudette was still at Microsoft and still held all his original jobs.

To meet Gates's goal of dominating world computing, Microsoft had to expand overseas. The company was already represented in Japan by ASCII, led by Kay Nishi. In Europe, operations were set up in the United Kingdom, France, and Germany, all under Scott Oki. Though Apple Computer didn't know it, Microsoft's international expansion was financed entirely with payments made by Apple to finance a special version of Microsoft's Multiplan spreadsheet program for the Apple IIe. Apple needed Multiplan because Lotus had refused to do a version of 1-2-3 for the IIe. Because Charles Simonyi had designed Multiplan to be very portable, moving it to the Apple IIe was easy, and the bulk of Apple's money was used to buy the world for Microsoft.

Even with real marketing and sales professionals finally on the job, accounting and computer systems in place, and looking every bit like a big company, Microsoft is still built around Bill Gates, and Bill Gates is still a nerd. During Microsoft's 1983 national sales meeting, which was held that year in Arizona, a group of company leaders, including Gates and Shirley, went for a walk in the desert to watch the sun set. Gates had been drinking and insisted on climbing up into the crook of a giant saguaro cactus. Shirley looked up at his new boss, who was squatting in the arms of the cactus, greasy hair plastered across his forehead, squinting at the setting sun.

"Someone get him down from there while he can still father children," Shirley ordered.

▸   ▸   ▸   ▸   ▸   ▸   ▸   ▸   ▸   ▸   ▸   ▸

CHAPTER FOURTEEN

COUNTER-REFORMATION

In Prudhoe Bay, in the oilfields of Alaska's North Slope, the sun goes down sometime in late November and doesn't appear again until January, and even then the days are so short that you can celebrate sunrise, high noon, and sunset all with the same cup of coffee. The whole day looks like that sliver of white at the base of your thumbnail.

It's cold in Prudhoe Bay in the wintertime, colder than I can say or you would believe—so cold that the folks who work for the oil companies start their cars around October and leave them running twenty-four hours a day clear through to April just so they won't freeze up.

Idling in the seemingly endless dark is not good for a car. Spark plugs foul and carburetors gum up. Gas mileage goes completely to hell, but that's okay; they've got the oil. Keeping those cars and trucks running night and pseudoday means that there are a lot of crummy, gas-guzzling, smoke-spewing vehicles in Prudhoe Bay in the winter, but at least they work.

Nobody ever lost his job for leaving a car running overnight during a winter in Prudhoe Bay.

And it used to be that nobody ever lost his job for buying computers from IBM.

But springtime eventually comes to Alaska. The tundra begins to melt, the days get longer than you can keep your eyes open, and the mosquitoes are suddenly thick as grass. It's time for an oil change and to give that car a rest. When the danger's gone—when the environment has improved to a point where any car can be counted on to make it through the night, when any tool could do the job—then efficiency and economy suddenly do become factors. At the end of June in Prudhoe Bay, you just might get in trouble for leaving a car running overnight, if there *was* a night, which there isn't.

IBM built its mainframe computer business on reliable service, not on computing performance or low prices. Whether it was in Prudhoe Bay or Houston, when the System 370/168 in accounting went down, IBM people were there *right now* to fix it and get the company back up and running. IBM customer hand holding built the most profitable corporation in the world. But when we're talking about a personal computer rather than a mainframe, and it's just one computer out of a dozen, or a hundred, or a thousand in the building, then having that guy in the white IBM coveralls standing by eventually stops being worth 30 percent or 50 percent more.

That's when it's springtime for IBM.

IBM's success in the personal computer business was a fluke. A company that was physically unable to invent anything in less than three years somehow produced a personal computer system and matching operating system in one year. Eighteen months later, IBM introduced the PC-XT, a marginally improved machine with a marginally improved operating system. Eighteen months after that, IBM introduced its real second-generation product, the PC-AT, with five times the performance of the XT.

From 1981 to 1984, IBM set the standard for personal computing and gave corporate America permission to take PCs seriously, literally creating the industry we know today. But after 1984, IBM lost control of the business.

Reality caught up with IBM's Entry Systems Division with the development of the PC-AT. From the AT on, it took IBM three years or better to produce each new line of computers. By mainframe standards, three years wasn't bad, but remember that mainframes are computers, while PCs are just piles of integrated circuits. PCs follow the price/performance curve for semiconductors, which says that performance has to double every eighteen months. IBM couldn't do that anymore. It should have been ready with a new line of industry-leading machines by 1986, but it wasn't. It was another company's turn.

Compaq Computer cloned the 8088-based IBM PC in a year and cloned the 80286-based PC-AT in six months. By 1986, IBM should have been introducing its 80386-based machine, but it didn't have one. Compaq couldn't wait for Big Blue and so went ahead and introduced its DeskPro 386. The 386s that soon followed from other clone makers were clones of the Compaq machine, not clones of IBM. Big Blue had fallen behind the performance curve and would never catch up. Let me say that a little louder: IBM WILL NEVER CATCH UP.

IBM had defined MS-DOS as the operating system of choice. It set a 16-bit bus standard for the PC-AT that determined how circuit cards from many vendors could be used in the same machine. These were benevolent standards from a market leader that needed the help of other hardware and software companies to increase its market penetration. That was all it took. Once IBM could no longer stay ahead of the performance curve, the IBM standards still acted as guidelines, so clone makers could take the lead from there, and they did. IBM saw its market share slowly start to fall.

But IBM was still the biggest player in the PC business, still had the the greatest potential for wreaking technical havoc, and knew better than any other company how to slow the game down to a more comfortable pace. Here are some market control techniques refined by Big Blue over the years.

Technique No. 1. Announce a direction, not a product. This is my favorite IBM technique because it is the most efficient one from Big Blue's perspective. Say the whole computer industry is waiting for IBM to come out with its next-generation machines, but instead the company makes a surprise announcement: "Sorry, no new computers this year, but that's because we are committing the company to move toward a family of computers based on gallium arsenide technology [or Josephson junctions, or optical computing, or even vegetable computing—it doesn't really matter]. Look for these powerful new computers in two years."

"Damn, I *knew* they were working on something big," say all of IBM's competitors as they scrap the computers they had been planning to compete with the derivative machines expected from IBM.

Whether IBM's rutabaga-based PC ever appears or not, all IBM competitors have to change their research and development focus, looking into broccoli and parsnip computing, just in case IBM is actually onto something. By stating a bold change of direction, IBM looks as if it's grasping the technical lead, when in fact all it's really doing is throwing competitors for a loop, burning up their R&D budgets, and ultimately making them wait up to two years for a new line of computers that may or may not ever appear. (IBM has been known, after all, to say later, "Oops, that just didn't work out," as they did with Josephson junction research.) And even when the direction is for real, the sheer market presence of IBM makes most other companies wait for Big Blue's machines to appear to see how they can make their own product lines fit with IBM's.

Whenever IBM makes one of these statements of direction, it's like the yellow flag coming out during an auto race. Everyone continues to drive, but nobody is allowed to pass.

IBM's Systems Application Architecture (SAA) announcement of 1987, which was supposed to bring a unified programming environment, user interface, and applications to most of its mainframe, minicomputer, and personal computer lines by 1989, was an example of such a statement of direction. SAA was for real, but major parts of it were still not ready in 1991.

Technique No. 2. Announce a real product, but do so long before you actually expect to deliver, disrupting the market for competitive products that are already shipping.

This is a twist on Technique No. 1 though aimed at computer buyers rather than computer builders. Because performance is always going up and prices are always going down, PC buyers love to delay purchases, waiting for something better. A major player like IBM can take advantage of this trend, using it to compete even when IBM doesn't yet have a product of its own to offer.

In the 1983–1985 time period, for example, Apple had the Lisa and the Macintosh, VisiCorp had VisiOn, its graphical computing environment for IBM PCs, Microsoft had shipped the first version of Windows, Digital Research produced GEM, and a little company in Santa Monica called Quarterdeck Office Systems came out with a product called DesQ. All of these products—even Windows, which came from Microsoft, IBM's PC software partner—were perceived as threats by IBM, which had no equivalent graphical product. To compete with these graphical environments that were already available, IBM announced its own software that would put pop-up windows on a PC screen and offer easy switching from application to application and data transfer from one program to another. The announcement came in the summer of 1984 at the same time the PC-AT was introduced. They called the new software TopView and said it would be available in about a year.

DesQ had been the hit of Comdex, the computer dealers' convention held in Atlanta in the spring of 1984. Just after the show, Quarterdeck raised $5.5 million in second-round venture funding, moved into new quarters just a block from the beach, and was happily shipping 2,000 copies of DesQ per month. DesQ had the advantage over most of the other windowing systems that it worked with existing MS-DOS applications. DesQ could run more than one application at a time, too—something none of the other systems (except Apple's Lisa) offered. Then IBM announced TopView. DesQ sales dropped to practically nothing, and the venture capitalists asked Quarterdeck for their money back.

All the potential DesQ buyers in the world decided in a single moment to wait for the truly incredible software IBM promised. They forgot, of course, that IBM was not particularly noted for incredible software—in fact, IBM had never developed PC software entirely on its own before. TopView was true Blue—written with no help from Microsoft.

The *idea* of TopView hurt all the other windowing systems and contributed to the death of VisiOn and DesQ. Quarterdeck dropped from fifty employees down to thirteen. Terry Myers, cofounder of Quarterdeck and one of the few women to run a PC software company, borrowed $20,000 from her mother to keep the company afloat while her programmers madly rewrote DesQ to be compatible with the yet-to-be-delivered TopView. They called the new program DesqView.

When TopView finally appeared in 1985, it was a failure. The product was slow and awkward to use, and it lived up to none of the promises IBM made. You can still buy TopView from IBM, but nobody does; it remains on the IBM product list strictly because removing it would require writing off all development expenses, which would hurt IBM's bottom line.

Technique No. 3. Don't announce a product, but do leak a few strategic hints, even if they aren't true.

IBM should have introduced a follow-on to the PC-AT in 1986 but it didn't. There were lots of rumors, sure, about a system generally referred to as the PC-2, but IBM staunchly refused to comment. Still, the PC-2 rumors continued, accompanied by sparse technical details of a machine that all the clone makers expected would include an Intel 80386 processor. And *maybe*, the rumors continued, the PC-2 would have a 32-bit bus, which would mean yet another technical standard for add-in circuit cards.

It would have been suicide for a clone maker to come out with a 386 machine with its own 32-bit bus in early 1986 if IBM was going to announce a similar product a month or three later, so the clone makers didn't introduce their new machines. They waited and waited for IBM to announce a new family of computers that never came. And during the time that Compaq and Dell, and AST, and the others were waiting for IBM to make its move, millions of PC-ATs were flowing into Fortune 1000 corporations, still bringing in the big bucks at a time when they shouldn't have still been viewed as top-of-the-line machines.

When Compaq Computer finally got tired of waiting and introduced its own DeskPro 386, it was careful to make its new machine use the 16-bit circuit cards intended for the PC-AT. Not even Compaq thought it could push a proprietary 32-bit bus standard in competition with IBM. The only 32-bit connections in the Compaq machine were between the processor and main memory; in every other respect, it was just like a 286.

Technique No. 4. Don't support anybody else's standards; make your own.

The original IBM Personal Computer used the PC-DOS operating system at a time when most other microcomputers used in business ran CP/M. The original IBM PC had a completely new bus standard, while nearly all of those CP/M machines used something called the S-100 bus. Pushing a new operating system and a new bus should have put IBM at a disadvantage, since there were thousands of CP/M applications and hundreds of S-100 circuit cards, and hardly any PC-DOS applications and less than half a dozen PC circuit cards available in 1981. But this was not just any computer start-up; this was *IBM*, and so what would normally have been a disadvantage became IBM's advantage. The IBM PC killed CP/M and the S-100 bus and gave Big Blue a full year with no PC-compatible competitors.

When the rest of the world did its computer networking with Ethernet, IBM invented another technology, called Token Ring. When the rest of the world thought that a multitasking workstation operating system meant Unix, IBM insisted on OS/2, counting on its influence and broad shoulders either to make the IBM standard a de facto standard or at least to interrupt the momentum of competitors.

Technique No. 5. Announce a product; then say you don't really mean it.

IBM has always had a problem with the idea of linking its personal computers together. PCs were cheaper than 3270 terminals, so IBM didn't want to make it too easy to connect PCs to its mainframes and risk hurting its computer terminal business. And linked PCs could, by sharing data, eventually compete with minicomputer or mainframe time-sharing systems, which were IBM's traditional bread and butter. Proposing an IBM standard for networking PCs or embracing someone else's networking standard was viewed in Armonk as a risky proposition. By the mid-1980s, though, other companies were already moving forward with plans to network IBM PCs, and Big Blue just couldn't stand the idea of all that money going into another company's pocket.

In 1985, then, IBM announced its first networking hardware and software for personal computers. The software was called the PC Network (later the PC LAN Program). The hardware was a circuit card that fit in each PC and linked them together over a coaxial cable, transferring data at up to 2 million bits per second. IBM sold $200 million worth of these circuit cards over the next couple of years. But that wasn't good enough (or bad enough) for IBM, which announced that the network cards, while they are a product, weren't part of an IBM *direction*. IBM's true networking direction was toward another hardware technology called Token Ring, which would be available, as I'm sure you can predict by now, *in a couple of years*.

Customers couldn't decide whether to buy the hardware that IBM was already selling or to wait for Token Ring, which would have higher performance. Customers who waited for Token Ring were punished for their loyalty, since IBM, which had the most advanced semiconductor plants in the world, somehow couldn't make enough Token Ring adapters to meet demand until well into 1990. The result was that IBM lost control of the PC networking business.

The company that absolutely controls the PC networking business is headquartered at the foot of a mountain range in Provo, Utah, just down the street from Brigham Young University. Novell Inc. runs the networking business today as completely as IBM ran the PC business in 1983. A lot of Novell's success has to do with the technical skills of those programmers who come to work straight out of BYU and who have no idea how much money they could be making in Silicon Valley. And a certain amount of its success can be traced directly to the company's darkest moment, when it was lucky enough to nearly go out of business in 1981.

Novell Data Systems, as it was called then, was a struggling maker of not very good CP/M computers. The failing company threw the last of its money behind a scheme to link its computers together so they could share a single hard disk drive. Hard disks were expensive then, and a California company, Corvus Systems, had already made a fortune linking Apple IIs together in a similar fashion. Novell hoped to do for CP/M computers what Corvus had done for the Apple II.

In September 1981, Novell hired three contract programmers to devise the new network hardware and software. Drew Major, Dale Neibaur, and Kyle Powell were techies who liked to work together and hired out as a unit under the name Superset. Superset—three guys who weren't even Novell employees—invented Novell's networking technology and still direct its development today. They still aren't Novell employees.

Companies like Ashton-Tate and Lotus Development ran into serious difficulties when they lost their architects. Novell and Microsoft, which have retained their technical leaders for over a decade, have avoided such problems.

In 1981, networking meant sharing a hard disk drive but not sharing data between microcomputers. Sure, your Apple II and my Apple II could be linked to the same Corvus 10-megabyte hard drive, but your data would be invisible to my computer. This was a safety feature, because the microcomputer operating systems of the time couldn't handle the concept of shared data.

Let's say I am reading the text file that contains your gothic romance just when you decide to add a juicy new scene to chapter 24. I am reading the file, adding occasional rude comments, when you grab the file and start to add text. Later, we both store the file, but which version gets stored: the one with my comments, or the one where Captain Phillips finally does the nasty with Lady Margaret? Who knows?

What CP/M lacked was a facility for directory locking, which would allow only one user at a time to change a file. I could read your romance, but if you were already adding text to it, directory locking would keep me from adding any comments. Directory locking could be used to make some data read only, and could make some data readable only by certain users. These were already important features in multiuser or networked systems but not needed in CP/M, which was written strictly for a single user.

The guys from Superset added directory locking to CP/M, they improved CP/M's mechanism for searching the disk directory, and they moved all of these functions from the networked microcomputer up to a specialized processor that was at the hard disk drive. By November 1981, they'd turned what was supposed to have been a disk server like Corvus's into a file server where users could share data. Novell's Data Management Computer could support twelve simultaneous users at the same performance level as a single-user CP/M system.

Superset, not Novell, decided to network the new IBM PC. The three hackers bought one of the first PCs in Utah and built the first PC network card. They did it all on their own and against the wishes of Novell, which just then finally ran out of money.

The venture capitalists whose money it was that Novell had used up came to Utah looking for salvageable technology and found only Superset's work worth continuing. While Novell was dismantled around them, the three contractors kept working and kept getting paid. They worked in isolation for two years, developing whole generations of product that were never sold to anyone.

The early versions of most software are so bad that good programmers usually want to throw them away but can't because ship dates have to be met. But Novell wasn't shipping anything in 1982–1983, so early versions of its network software were thrown away and started over again. Novell was able take the time needed to come up with the correct architecture, a rare luxury for a start-up, and subsequently the company's greatest advantage. Going broke turned out to have been very good for Novell.

Novell hardware was so bad that the company concentrated almost completely on software after it started back in business in 1983. All the other networking companies were trying to sell hardware. Corvus was trying to sell hard disks. Televideo was trying to sell CP/M boxes. 3Com was trying to sell Ethernet network adapter cards. None of these companies saw any advantage to selling its software to go with another company's hard disk, computer, or adapter card. They saw all the value in the hardware, while Novell, which had lousy hardware and knew it, decided to concentrate on networking software that would work with *every* hard drive, *every* PC, and *every* network card.

By this time Novell had a new leader in Ray Noorda, who'd bumped through a number of engineering, then later marketing and sales, jobs in the minicomputer business. Noorda saw that Novell's value lay in its software. By making wiring a nonissue, with Novell's software—now called Netware—able to run on any type of networking scheme, Noorda figured it would be possible to stimulate the next stage of growth. "Growing the market" became Noorda's motto, and toward that end he got Novell back in the hardware business but sold workstations and network cards literally at cost just to make it cheaper and easier for companies to decide to network their offices. Ray Noorda was not a popular man in Silicon Valley.

In 1983, when Noorda was taking charge of Novell, IBM asked Microsoft to write some PC networking software. Microsoft knew very little about networking in 1983, but Bill Gates was not about to send his major customer away, so Microsoft got into the networking business.

"Our networking effort wasn't serious until we hired Darryl Rubin, our network architect," admitted Microsoft's Steve Ballmer in 1991.

Wait a minute, Steve, did anyone tell IBM back in 1983 that Microsoft wasn't really serious about this networking stuff? Of course not.

Like most of Microsoft's other stabs at new technology, PC networking began as a preemptive strike rather than an actual product. The point of Gates's agreeing to do IBM's network software was to keep IBM as a customer, not to do a good product. In fact, Microsoft's entry into most new technologies follows this same plan, with the first effort being a preemptive strike, the second effort being market research to see what customers really want in a product, and the third try is the real product. It happened that way with Microsoft's efforts at networking, word processing, and Windows, and will continue in the company's current efforts in multimedia and pen-based computing. It's too bad, of course, that hundreds of thousands of customers spend millions and millions of dollars on those early efforts—the ones that aren't real products. But heck, that's their problem, right?

Microsoft decided to build its network technology on top of DOS because that was the company franchise. All new technologies were conceived as extensions to DOS, keeping the old technology competitive—or at least looking so—in an increasingly complex market. But DOS wasn't a very good system on which to build a network operating system. DOS was limited to 640K of memory. DOS had an awkward file structure that got slower and slower as the number of files increased, which could become a major problem on a server with thousands of files. In contrast, Novell's Netware could use megabytes of memory and had a lightning-fast file system. After all, Netware was built from scratch to be a network operating system, while Microsoft's product wasn't.

MS-Net appeared in 1985. It was licensed to more than thirty different hardware companies in the same way that MS-DOS was licensed to makers of PC clones. Only three versions of MS-Net actually appeared, including IBM's PC LAN program, a dog.

The final nail in Microsoft's networking coffin was also driven in 1985 when Novell introduced Netware 2.0, which ran on the 80286 processor in IBM's PC-AT. You could run MS-Net on an AT also but only in the mode that emulated an 8086 processor and was limited to addressing 640K. But Netware on an AT took full advantage of the 80286 and could address up to 16 megabytes of RAM, making Novell's software vastly more powerful than Microsoft's.

This business of taking software written for the 8086 processor and porting it to the 80286 normally required completely rewriting the software by hand, often taking years of painstaking effort. It wasn't just a matter of recompiling the software, of having a machine do the translation, because Microsoft staunchly maintained that there was no way to recompile 8086 code to run on an 80286. Bill Gates swore that such a recompile was impossible. But Drew Major of Superset didn't know what Bill Gates knew, and so he figured out a way to recompile 8086 code to run on an 80286. What should have taken months or years of labor was finished in a week, and Novell had won the networking war. Six years and more than $100 million later, Microsoft finally admitted defeat.

≈≈

Meanwhile, back in Boca Raton, IBM was still struggling to produce a follow-on to the PC-AT. The reason that it began taking

IBM so long to produce new PC products was the difference between strategy and tactics. Building the original IBM PC was a tactical exercise designed to test a potential new market by getting a product out as quickly as possible. But when the new market turned out to be ten times larger than anyone at IBM had realized and began to affect the sales of other divisions of the company, PCs suddenly became a strategic issue. And strategy takes time to develop, especially at IBM.

Remember that there is nobody working at IBM today who recalls those sun-filled company picnics in Endicott, New York, back when the company was still small, the entire R&D department could participate in one three-legged race, and inertia was not yet a virtue. The folks who work at IBM today generally like the fact that it is big, slow moving, and safe. IBM has built an empire by moving deliberately and hiring third-wave people. Even Don Estridge, who led the tactical PC effort up through the PC-AT, wasn't welcome in a strategic personal computer operation; Estridge was a second-wave guy at heart and so couldn't be trusted. That's why Estridge was promoted into obscurity, and Bill Lowe, who'd proved that he was a company man, a true third waver with only occasional second-wave leanings that could, and were, beaten out of him over time, was brought back to run the PC operations.

As an enormous corporation that had finally decided personal computers were part of its strategic plan, IBM laboriously reexamined the whole operation and started funding backup ventures to keep the company from being too dependent on any single PC product development effort. Several families of new computers were designed and considered, as were at least a couple of new operating systems. All of this development and deliberation takes time.

Even the vital relationship with Bill Gates was reconsidered in 1985, when IBM thought of dropping Microsoft and DOS altogether in favor of a completely new operating system. The idea was to port to the Intel 286 processor operating system software from a California company called Metaphor Computer Systems. The Metaphor software was yet another outgrowth of work done at Xerox PARC and ran then strictly on IBM mainframes, offering an advanced office automation system with a graphical user interface. The big corporate users who were daring enough to try Metaphor loved it, and IBM dreamed that converting the software to run on PCs would draw personal computers seamlessly into the mainframe world in a way that wouldn't be so directly competitive with its other product lines. Porting Metaphor software would also have brought IBM a major role in application software for its PCs—an area where the company had so far failed.

Since Microsoft wasn't even supposed to know that this Metaphor experiment was happening, IBM chose Lotus Development to port the software. The programmers at Lotus had never written an operating system, but they knew plenty about Intel processor architecture, since the high performance of Lotus 1-2-3 came mainly from writing directly to the processor, avoiding MS-DOS as much as possible.

Nothing ever came of the Lotus/Metaphor operating system, which turned out to be an IBM fantasy. Technically, it was asking too much of the 80286 processor. The 80386 might have handled the job, but for other strategic reasons, IBM was reluctant to move up to the 386.

IBM has had a lot of such fantasies and done a lot of negotiating and investigating whacko joint ventures with many different potential software partners. It's a way of life at the largest computer company in the world, where keeping on top of the industry is accomplished through just this sort of diplomacy. Think of dogs sniffing each other.

IBM couldn't go forever without replacing the PC-AT, and eventually it introduced a whole new family of microcomputers in April 1987. These were the Personal System/2s and came in four flavors: Models 30, 50, 60, and 80. The Model 30 used an 8086 processor, the Models 50 and 60 used an 80286, and the Model 80 was IBM's first attempt at an 80386-based PC. The 286 and 386 machines used a new bus standard called the Micro Channel, and all of the PS/2s had 3.5-inch floppy disk drives. By changing hardware designs, IBM was again trying to have the market all to itself.

A new bus standard meant that circuit cards built for the IBM PC, XT, or AT models wouldn't work in the PS/2s, but the new bus, which was 32 bits wide, was supposed to offer so much higher performance that a little more cost and inconvenience would be well worthwhile. The Micro Channel was designed by an iconoclastic (by IBM standards) engineer named Chet Heath and was reputed to beat the shit out of the old 16-bit AT bus. It was promoted as the next generation of personal computing, and IBM expected the world to switch to its Micro Channel in just the way it had switched to the AT bus in 1984.

But when we tested the PS/2s at *InfoWorld*, the performance wasn't there. The new machines weren't even as fast as many AT clones. The problem wasn't the Micro Channel; it was IBM. Trying to come up with a clever work-around for the problem of generating a new product line every eighteen months when your organization inherently takes three years to do the job, product planners in IBM's Entry Systems Division simply decided that the first PS/2s would use only half of the features of the Micro Channel bus. The company deliberately shipped hobbled products so that, eighteen months later, it could *discover* all sorts of neat additional Micro Channel horsepower, which would be presented in a whole new family of machines using what would then be called Micro Channel 2.

IBM screwed up in its approach to the Micro Channel. Had it introduced the whole product in 1987, doubling the performance of competitive hardware, buyers would have followed IBM to the new standard as they had before. They could have led the industry to a new 32-bit bus standard—one where IBM again would have had a technical advantage for a while. But instead, Big Blue held back features and then tried to scare away clone makers by threatening legal action and talking about granting licenses for the new bus only if liccnsees paid 5 percent royalties on both their new Micro Channel clones and *on every PC, XT, or AT clone they had ever built*. The only result of this new hardball attitude was that an industry that had had little success defining a new bus standard by itself was suddenly solidified against IBM. Compaq Computer led a group of nine clone makers that defined their own 32-bit bus standard in competition with the Micro Channel. Compaq led the new group, but IBM made it happen.

From IBM's perspective, though, its approach to the Micro Channel and the PS/2s was perfectly correct since it acted to protect Big Blue's core mainframe and minicomputer products. Until very recently, IBM concentrated more on the threat that PCs posed to its larger computers than on the opportunities to sell ever more millions of PCs. Into the late 1980s, IBM still saw itself primarily as a maker of large computers.

Along with new PS/2 hardware, IBM announced in 1987 a new operating system called OS/2, which had been under development at Microsoft when IBM was talking with Metaphor and Lotus. The good part about OS/2 was that it was a true multitasking operating system that allowed several programs to run at the same time on one computer. The bad part about OS/2 was that it was designed by IBM.

When Bill Lowe sent his lieutenants to Microsoft looking for an operating system for the IBM PC, they didn't carry a list of specifications for the system software. They were looking for something that was ready—software they could just slap on the new machine and run. And that's what Microsoft gave IBM in PC-DOS: an off-the-shelf operating system that would run on the new hardware. *Microsoft,* not IBM, decided what DOS would look like and act like. DOS was a Microsoft product, not an IBM product, and subsequent versions, though they appeared each time in the company of new IBM hardware, continued to be 100 percent Microsoft code.

OS/2 was different. OS/2 was strategic, which meant that it was too important to be left to the design whims of Microsoft alone. OS/2 would be designed by IBM and just coded by Microsoft. Big mistake.

OS/2 1.0 was designed to run on the 80286 processor. Bill Gates urged IBM to go straight for the 80386 processor as the target for OS/2, but IBM was afraid that the 386 would offer performance too close to that of its minicomputers. Why buy an AS/400 minicomputer for $200,000, when half a dozen networked PS/2 Model 80s running OS/2-386 could give twice the performance for one third the price? The only reason IBM even developed the 386-based Model 80, in fact, was that Compaq was already selling thousands of its DeskPro 386s. Over the objections of Microsoft, then, OS/2 was aimed at the 286, a chip that Gates correctly called "brain damaged."

OS/2 had both a large address space and virtual memory. It had more graphics options than either Windows or the Macintosh, as well as being multithreaded and multitasking. OS/2 looked terrific on paper. But what the paper didn't show was what Gates called "poor code, poor design, poor process, and other overhead" thrust on Microsoft by IBM.

While Microsoft retained the right to sell OS/2 to other computer makers, this time around IBM had its own special version of OS/2, Extended Edition, which included a database called the

Data Manager, and an interface to IBM mainframes called the Communication Manager. These special extras were intended to tie OS/2 and the PS/2s into their true function as very smart mainframe terminals. IBM had much more than competing with Compaq in mind when it designed the PS/2s. IBM was aiming toward a true counterreformation in personal computing, leading millions of loyal corporate users back toward the holy mother church—the mainframe.

IBM's dream for the PS/2s, and for OS/2, was to play a role in leading American business *away* from the desktop and back to big expensive computers. This was the objective of SAA—IBM's plan to integrate its personal computers and mainframes—and of what they hoped would be SAA's compelling application, called OfficeVision.

On May 16, 1989, I sat in an auditorium on the ground floor of the IBM building at 540 Madison Avenue. It was a rainy Tuesday morning in New York, and the room, which was filled with bright television lights as well as people, soon took on the distinctive smell of wet wool. At the front of the room stood a podium and a long table, behind which sat the usual IBM suspects—a dozen conservatively dressed, overweight, middle-aged white men.

George Conrades, IBM's head of U.S. marketing, appeared behind the podium. Conrades, 43, was on the fast career track at IBM. He was younger than nearly all the other men of IBM who sat at the long table behind him, waiting to play their supporting roles. Behind the television camera lens, 25,000 IBM employees, suppliers, and key customers spread across the world watched the presentation by satellite.

The object of all this attention was a computer software product from IBM called OfficeVision, the result of 4,000 man-years of effort at a cost of more than a billion dollars.

To hear Conrades and the others describe it through their carefully scripted performances, OfficeVision would revolution-ize American business. Its "programmable terminals" (PCs) with their immense memory and processing power would gather data from mainframe computers across the building or across the planet, seeking out data without users' having even to know where the data were stored and then compiling them into color-ful and easy-to-understand displays. OfficeVision would bring top executives for the first time into intimate—even casual—contact with the vital data stored in their corporate computers. Beyond the executive suite, it would offer access to data, sophisticated communication tools, and intuitive ways of viewing and using information throughout the organization. OfficeVision would even make it easier for typists to type and for file clerks to file.

In the glowing words of Conrades, OfficeVision would make American business more competitive and more profitable. If the experts were right that computing would determine the future success or failure of American business, then OfficeVision simply was that future. It would make that success.

"And all for an average of $7,600 per desk," Conrades said, "not including the IBM mainframe computers, of course."

The truth behind this exercise in worsted wool and public relations is that OfficeVision was not at all the future of comput-ing but rather its past, spruced up, given a new coat of paint, and trotted out as an all-new model when, in fact, it was not new at all. In the eyes of IBM executives and their strategic partners, though, OfficeVision had the appearance of being new, which was even better. To IBM and the world of mainframe computers, danger lies in things that are truly new.

With its PS/2s and OS/2 and OfficeVision, IBM was trying to get a jump on a new wave of computing that everyone knew was on its way. The first wave of computing was the mainframe. The second wave was the minicomputer. The third wave was the PC.

Now the fourth wave—generally called network computing—seemed imminent, and IBM's big-bucks commitment to SAA and to OfficeVision was its effort to make the fourth wave look as much as possible like the first three. Mainframes would do the work in big companies, minicomputers in medium-sized companies, and PCs would serve small business as well as acting as "programmable terminals" for the big boys with their OfficeVision setups.

Sadly for IBM, by 1991, OfficeVision still hadn't appeared, having tripped over mountains of bad code, missed delivery schedules, and facing the fact of life that corporate America is only willing to invest less than 10 percent of each worker's total compensation in computing resources for that worker. That's why secretaries get $3,000 PCs and design engineers get $10,000 workstations. OfficeVision would have cost at least double that amount per desk, had it worked at all, so today IBM is talking about a new, slimmed-down OfficeVision 2.0, which will probably fail too.

When OS/2 1.0 finally shipped months after the PS/2 introduction, every big shot in the PC industry asked his or her market research analysts when OS/2 unit sales would surpass sales of MS-DOS. The general consensus of analysts was that the crossover would take place in the early 1990s, perhaps as soon as 1991. It didn't happen.

Time to talk about the realities of market research in the PC industry. Market research firms make surveys of buyers and sellers, trying to predict the future. They gather and sift through millions of bytes of data and then apply their S-shaped demand curves, predicting what will and won't be a hit. Most of what they do is voodoo. And like voodoo, whether their work is successful depends on the state of mind of their victim/customer.

Market research customers are hardware and software companies paying thousands—sometimes hundreds of thousands—of dollars, primarily to have their own hunches confirmed. Remember that the question on everyone's mind was when unit sales of OS/2 would exceed those of DOS. Forget that OS/2 1.0 was late. Forget that there was no compelling application for OS/2. Forget that the operating system, when it did finally appear, was buggy as hell and probably shouldn't have been released at all. Forget all that, and think only of the question, which was: *When* will unit sales of OS/2 exceed those of DOS? The assumption (and the flaw) built into this exercise is that OS/2, because it was being pushed by IBM, was destined to overtake DOS, which it hasn't. But given that the paying customers wanted OS/2 to succeed and that the research question itself suggested that OS/2 would succeed, market research companies like Dataquest, InfoCorp, and International Data Corporation dutifully crazy-glued their usual demand curves on a chart and predicted that OS/2 would be a big hit. There were no dissenting voices. Not a single market research report that I read or read about at that time predicted that OS/2 would be a failure.

Market research firms tend to serve the same function for the PC industry that a lamppost does for a drunk.

OS/2 1.0 was a dismal failure. Sales were pitiful. Performance was pitiful, too, at least in that first version. Users didn't need OS/2 since they could already multitask their existing DOS applications using products like Quarterdeck's DesqView. Independent software vendors, who were attracted to OS/2 by the lure of IBM, soon stopped their OS/2 development efforts as the operating system's failure became obvious. But the failure of OS/2 wasn't all IBM's fault. Half of the blame has to go on the computer memory crisis of the late 1980s.

OS/2 made it possible for PCs to access far more memory than the pitiful 640K available under MS-DOS. On a 286 machine,

OS/2 could use up to 16 megabytes of memory and in fact seemed to require at least 4 megabytes to perform acceptably. Alas, this sudden need for six times the memory came at a time when American manufacturers had just abandoned the dynamic random-access memory (DRAM) business to the Japanese.

In 1975, Japan's Ministry for International Trade and Industry had organized Japan's leading chip makers into two groups—NEC-Toshiba and Fujitsu-Hitachi-Mitsubishi—to challenge the United States for the 64K DRAM business. They won. By 1985, these two groups had 90 percent of the U.S. market for DRAMs. American companies like Intel, which had started out in the DRAM business, quit making the chips because they weren't profitable, cutting world DRAM production capacity as they retired. Then, to make matters worse, the United States Department of Commerce accused the Asian DRAM makers of dumping —selling their memory chips in America at less than what it cost to produce them. The Japanese companies cut a deal with the United States government that restricted their DRAM distribution in America—at a time when we had no other reliable DRAM sources. Big mistake. Memory supplies dropped just as memory demand rose, and the classic supply-demand effect was an increase in DRAM prices, which more than doubled in a few months. Toshiba, which was nearly the only company making 1 megabit DRAM chips for a while, earned more than $1 billion in profits on its DRAM business in 1989, in large part because of the United States government.

Doubled prices are a problem in any industry, but in an industry based on the idea of prices' continually dropping, such an increase can lead to panic, as it did in the case of OS/2. The DRAM price bubble was just that—a bubble—but it looked for a while like the end of the world. Software developers who were already working on OS/2 projects began to wonder how many users would be willing to invest the $1,000 that it was suddenly costing to add enough memory to their systems to run OS/2. Just as raising prices killed demand for Apple's Macintosh in the fall of 1988 (Apple's primary reason for raising prices was the high cost of DRAM), rising memory prices killed both the supply and demand for OS/2 software.

Then Bill Gates went into seclusion for a week and came out with the sudden understanding that DOS was good for Microsoft, while OS/2 was probably bad. Annual reading weeks, when Gates stays home and reads technical reports for seven days straight and then emerges to reposition the company, are a tradition at Microsoft. Nothing is allowed to get in the way of planned reading for Chairman Bill. During one business trip to South America, for example, the head of Microsoft's Brazilian operation tried to impress the boss by taking Gates and several women yachting for the weekend. But this particular weekend had been scheduled for reading, so Bill, who is normally very much on the make, stayed below deck reading the whole time.

Microsoft had loyally followed IBM in the direction of OS/2. But there must have been an idea nagging in the back of Bill Gates's mind. By taking this quantum leap to OS/2, IBM was telling the world that DOS was dead. If Microsoft followed IBM too closely in this OS/2 campaign, it was risking the more than $100 million in profits generated each year by DOS—profits that mostly didn't come from IBM. During one of his reading weeks, Gates began to think about what he called "DOS as an asset" and in the process set Microsoft on a collision course with IBM.

Up to 1989, Microsoft followed IBM's lead, dedicating itself publicly to OS/2 and promising versions of all its major applications that would run under the new operating system. On the surface, all was well between Microsoft and IBM. Under the surface, there were major problems with the relationship. A feisty (for IBM) band of graphics programmers at IBM's lab in Hursley,

England, first forced Microsoft to use an inferior and difficult-to-implement graphics imaging model in Presentation Manager and then later committed all the SAA operating systems, including OS/2, to using PostScript, from the hated house of Warnock—Adobe Systems.

Although by early 1990, OS/2 was up to version 1.2, which included a new file system and other improvements, more than 200 copies of DOS were still being sold for every copy of OS/2. Gates again proposed to IBM that they abandon the 286-based OS/2 product entirely in favor of a 386-based version 2.0. Instead, IBM's Austin, Texas, lab whipped up its own OS/2 version 1.3, generally referred to as OS/2 Lite. Outwardly, OS/2 1.3 tasted great and was less filling; it ran much faster than OS/2 1.2 and required only 2 megabytes of memory. But OS/2 1.3 sacrificed subsystem performance to improve the speed of its user interface, which meant that it was not really as good a product as it appeared to be. Thrilled finally to produce some software that was well received by reviewers, IBM started talking about basing all its OS/2 products on 1.3—even its networking and database software, which didn't even *have* user interfaces that needed optimizing. To Microsoft, which was well along on OS/2 2.0, the move seemed brain damaged, and this time they said so.

Microsoft began moving away from OS/2 in 1989 when it became clear that DOS wasn't going away, nor was it in Microsoft's interest for it to go away. The best solution for Microsoft would be to put a new face on DOS, and that new face would be yet another version of Windows. Windows 3.0 would include all that Microsoft had learned about graphical user interfaces from seven years of working on Macintosh applications. Windows 3.0 would also be aimed at more powerful PCs using 386 processors—the PCs that Bill Gates expected to dominate business desktops for most of the 1990s. Windows would preserve DOS's asset value for Microsoft and would give users

90 percent of the features of OS/2, which Gates began to see more and more as an operating system for network file servers, database servers, and other back-end network applications that were practically invisible to users.

IBM wanted to take from Microsoft the job of defining to the world what a PC operating system was. Big Blue wanted to abandon DOS in favor of OS/2 1.3, which it thought could be tied more directly into IBM hardware and applications, cutting out the clone makers in the process. Gates thought this was a bad idea that was bound to fail. He recognized, even if IBM didn't, that the market had grown to the point where no one company could define and defend an operating system standard by itself. Without Microsoft's help, Gates thought IBM would fail. With *IBM's* help, which Gates viewed more as meddling than assistance, *Microsoft* might fail. Time for a divorce.

Microsoft programmers deliberately slowed their work on OS/2 and especially on Presentation Manager, its graphical user interface. "What incentive does Microsoft have to get [OS/2-PM] out the door before Windows 3?" Gates asked two marketers from Lotus over dinner following the 1990 Computer Bowl trivia match in April 1990. "Besides, six months after Windows 3 ships it will have greater market share than PM will *ever* have. OS/2 applications won't have a chance."

Later that night over drinks, Gates speculated that IBM would "fold" in seven years, though it could last as long as ten or twelve years if it did everything right. Inevitably, though, IBM would die, and Bill Gates was determined that Microsoft would not go down too.

The loyal Lotus marketers prepared a seven-page memo about their inebriated evening with Chairman Bill, giving copies of it to their top management. Somehow I got a copy of the memo, too. And a copy eventually landed on the desk of IBM's Jim Cannavino, who had taken over Big Blue's PC operations from Bill Lowe. The end was near for IBM's special relationship with Microsoft.

Over the course of several months in 1990, IBM and Microsoft negotiated an agreement leaving DOS and Windows with Microsoft and OS/2 1.3 and 2.0 with IBM. Microsoft's only connection to OS/2 was the right to develop version 3.0, which would run on non-Intel processors and might not even share all the features of earlier versions of OS/2.

The Presentation Manager programmers in Redmond, who had been having Nerfball fights with their Windows counterparts every night for months, suddenly found themselves melded into the Windows operation. A cross-licensing agreement between the two companies remained in force, allowing IBM to offer subsequent versions of DOS to its customers and Microsoft the right to sell versions of OS/2, but the emphasis in Redmond was clearly on DOS and Windows, not OS/2.

"Our strategy for the 90's is Windows—one evolving architecture, a couple of implementations," Bill Gates wrote. "Everything we do should focus on making Windows more successful."

Windows 3.0 was introduced in May 1990 and sold more than 3 million copies in its first year. Like many other Microsoft products, this third try was finally the real thing. And since it had a head start over its competitors in developing applications that could take full advantage of Windows 3.0, Microsoft was more firmly entrenched than ever as the number one PC software company, while IBM struggled for a new identity. All those other software developers, the ones who had believed three years of Microsoft and IBM predictions that OS/2's Presentation Manager was the way to go, quickly shifted their OS/2 programmers over to writing Windows applications.

CHAPTER FIFTEEN

FUTURE COMPUTING

Remember *Pogo*? Pogo was *Doonesbury* in a swamp, the first politi-cal cartoon good enough to make it off the editorial page and into the high-rent district next to the horoscope. Pogo was a 'possum who looked as if he was dressed for a Harvard class reunion and who acted as the moral conscience for the first generation of Americans who knew how to read but had decided not to.

The *Pogo* strip remembered by everyone who knows what the heck I am even talking about is the one in which the little 'possum says, "We have met the enemy and he is us." But today's sermon is based on the line that follows in the next panel of that strip—a line that hardly anyone remembers. He said, "We are surrounded by insurmountable opportunity."

We *are* surrounded by insurmountable opportunity.

Fifteen years ago, a few clever young people invented a type of computer that was so small you could put it on a desk and so useful and cheap to own that America found places for more than 60 million of them. These same young people also invented games to play on those computers and business applications that were so powerful and so useful that we nearly all became com-puter literate, whether we wanted to or not.

Remember computer literacy? We were all supposed to become computer literate, or something terrible was going to happen to America. Computer literacy meant knowing how to program a computer, but that was before we really had an idea what personal computers could be used for. Once people had a reason for using computers other than to learn *how* to use computers, we stopped worrying about computer literacy and got on with our spreadsheets.

And that's where we pretty much stopped.

There is no real difference between an Apple II running Visi-Calc and an IBM PS/2 Model 70 running Lotus 1-2-3 version 3.0. Sure, the IBM has 100 times the speed and 1,000 times the storage of the Apple, but they are both just spreadsheet machines. Put the same formulas in the same cells, and both machines will give the same answer.

In 1984, marketing folks at Lotus tried to contact the people who bought the first ten copies of VisiCalc in 1979. Two users could not be reached, two were no longer using computers at all, three were using Lotus 1-2-3, and three were still using VisiCalc on their old Apple IIs. Those last three people were still having their needs met by a five-year-old product.

Marketing is the stimulation of long-term demand by solving customer problems. In the personal computer business, we've been solving more or less the same problem for at least ten years. Hardware is faster and software is more sophisticated, but the only real technical advances in software in the last ten years have been the Lisa's multitasking operating system and graphical user interface, Adobe's PostScript printing technology, and the ability to link users together in local area networks.

Ken Okin, who was in charge of hardware engineering for the Lisa and now heads the group designing Sun Microsystems' newest workstations, keeps a Lisa in his office at Sun just to help his people put their work in perspective. "We still have a multitasking operating system with a graphical user interface and bit-mapped screen, but back then we did it with half a mip [one mip equals one million computer instructions per second] in 1 megabyte of RAM," he said. "Today on my desk I have basically the same system, but this time I have 16 mips and an editor that doesn't seem to run in anything less than 20 megabytes of RAM. It runs faster, sure, but what will it do that is different from the Lisa? It can do round windows; that's all I can find that's new. *Round windows*, great!"

There hasn't been much progress in software for two reasons. The bigger reason is that companies like Microsoft and Lotus have been making plenty of money introducing more and more people to essentially the same old software, so they saw little reason to take risks on radical new technologies. The second reason is that radical new software technologies seem to require equally radical increases in hardware performance, something that is only now starting to take place as 80386- and 68030-based computers become the norm.

Fortunately for users and unfortunately for many companies in the PC business, we are about to break out of the doldrums of personal computing. There is a major shift happening right now that is forcing change on the business. Four major trends are about to shift PC users into warpspeed: standards-based computing, RISC processors, advanced semiconductors, and the death of the mainframe. Hold on!

In the early days of railroading in America, there was no rule that said how far apart the rails were supposed to be, so at first every railroad set its rails a different distance apart, with the result that while a load of grain could be sent from one part of the country to another, the car it was loaded in couldn't be. It took about thirty years for the railroad industry to standardize on just a couple of gauges of track. As happens in this business, one type of track, called *standard gauge,* took about 85 percent of the market.

A standard gauge is coming to computing, because no one company—even IBM—is powerful enough to impose its way of doing things on all the other companies. From now on, successful computers and software will come from companies that build them from scratch with the idea of working with computers and software made by their competitors. This heretical idea was foisted on us all by a company called Sun Microsystems, which invented the whole concept of open systems computing and has grown into a $4 billion company literally by giving software away.

Like nearly every other venture in this business, Sun got its start because of a Xerox mistake. The Defense Advanced Research Projects Agency wanted to buy Alto workstations, but the Special Programs Group at Xerox, seeing a chance to stick the feds for the entire Alto development budget, marked up the price too high even for DARPA. So DARPA went down the street to Stanford University, where they found a generic workstation based on the Motorola 68000 processor. Designed originally to run on the Stanford University Network, it was called the S.U.N. workstation.

Andy Bechtolscheim, a Stanford graduate student from Germany, had designed the S.U.N. workstation, and since Stanford was not in the business of building computers for sale any more than Xerox was, he tried to interest established computer companies in filling the DARPA order. Bob Metcalfe at 3Com had a chance to build the S.U.N. workstation but turned it down. Bechtolscheim even approached IBM, borrowing a tuxedo from the Stanford drama department to wear for his presentation because his friends told him Big Blue was a very formal operation.

He appeared at IBM wearing the tux, along with a tastefully contrasting pair of white tennis shoes. For some reason, IBM decided not to build the S.U.N. workstation either.

Since all the real computer companies were uninterested in building S.U.N. workstations, Bechtolscheim started his own company, Sun Microsystems. His partners were Vinod Khosla and Scott McNealy, also Stanford grad students, and Bill Joy, who came from Berkeley. The Stanford contingent came up with the hardware design and a business plan, while Joy, who had played a major role in writing a version of the Unix operating system at Berkeley, was Mr. Software.

Sun couldn't afford to develop proprietary technology, so it didn't develop any. The workstation design itself was so bland that Stanford University couldn't find any basis for demanding royalties from the start-up. For networking they embraced Bob Metcalfe's Ethernet, and for storage they used off-the-shelf hard disk drives built around the Small Computer System Interface (SCSI) specification. For software, they used Bill Joy's Berkeley Unix. Berkeley Unix worked well on a VAX, so Bechtolscheim and friends just threw away the VAX and replaced it with cheaper hardware. The languages, operating system, networking, and windowing systems were all standard.

Sun learned to establish de facto standards by giving source code away. It was a novel idea, born of the Berkeley Unix community, and rather in keeping with the idea that for some boys, a girl's attractiveness is directly proportional to her availability. For example, Sun virtually gave away licenses for its Network Filing System networking scheme, which had lots of bugs and some severe security problems, but it was free and so became a de facto standard virtually overnight. Even IBM licensed NFS. This giving away of source code allowed Sun to succeed, first by being the standard setter and then following up with the first hardware to support that standard.

By 1985, Sun had defined a new category of computer, the engineering workstation, but competitors were starting to catch on and catch up to Sun. The way to remain ahead of the industry, they decided, was to increase performance steadily, which they could do by using a RISC processor—except that there weren't any RISC processors for sale in 1985.

RISC is an old IBM idea called Reduced Instruction Set Computing. RISC processors were incredibly fast devices that gained their speed from a simple internal architecture that implements only a few computer instructions. Where a Complex Instruction Set Computer (CISC) might have a special "walk across the room but don't step on the dog" instruction, RISC processors can usually get faster performance by using several simpler instructions: walk–walk–step over–walk–walk.

RISC processors are cheaper to build because they are smaller and more can be fit on one piece of silicon. And because they have fewer transistors (often under 100,000), yields are higher too. It's easier to increase the clock speed of RISC chips, making them faster. It's easier to move RISC designs from one semiconductor technology to a faster one. And because RISC forces both hardware and software designers to keep it simple, stupid, they tend to be more robust.

Sun couldn't interest Intel or Motorola in doing one. Neither company wanted to endanger its lucrative CISC processor business. So Bill Joy and Dave Patterson designed a processor of their own in 1985, called SPARC. By this time, both Intel and Motorola had stopped allowing other semiconductor companies to license their processor designs, thus keeping all the high-margin sales in Santa Clara and Schaumberg, Illinois. This, of course, pissed off the traditional second source manufacturers, so Sun signed up those companies to do SPARC.

Since Sun designed the SPARC processor, it could buy them more cheaply than any other computer maker. Sun engineers knew, too, when higher-performance versions of the SPARC were going to be introduced. These facts of life have allowed Sun to dominate the engineering workstation market, as well as making important inroads into other markets formerly dominated by IBM and DEC.

Sun scares hardware and software competitors alike. The company practically gives away system software, which scares companies like Microsoft and Adobe that prefer to sell it. The industry is abuzz with software consortia set up with the intention to do better standards-based software than Sun does but to sell it, not give it away.

Sun also scares entrenched hardware competitors like DEC and IBM by actually encouraging cloning of its hardware architecture, relying on a balls-to-the-wall attitude that says Sun will stay in the high-margin leading edge of the product wave simply by bringing newer, more powerful SPARC systems to market sooner than any of its competitors can.

DEC has tried, and so far failed, to compete with Sun, using a RISC processor built by MIPS Computer Systems. Figuring if you can't beat them, join them, H-P has actually allied with Sun to do software. IBM reacted to Sun by building a RISC processor of its own too. Big Blue spent more on developing its Sun killer, the RS/6000, than it would have cost to buy Sun Microsystems outright. The RS/6000, too, is a relative failure.

≈≈

Why did Bill Gates, in his fourth consecutive hour of sitting in a hotel bar in Boston, sinking ever deeper into his chair, tell the marketing kids from Lotus Development that IBM would be out of business in seven years? What does Bill Gates know that we don't know?

Bill Gates knows that the future of computing will unfold on desktops, not in mainframe computer rooms. He knows that IBM has not had a very good handle on the desktop software market. He *thinks* that without the assistance of Microsoft, IBM will eventually forfeit what advantage it currently has in personal computers.

Bill Gates is a smart guy.

But you and I can go even further. We can predict the date by which the old IBM—IBM the mainframe computing giant—will be dead. We can predict the very day that the mainframe computer era will end.

Mainframe computing will die with the coming of the millennium. On December 31, 1999, right at midnight, when the big ball drops and people are kissing in New York's Times Square, the era of mainframe computing will be over.

Mainframe computing will end that night because a lot of people a long time ago made a simple mistake. Beginning in the 1950s, they wrote inventory programs and payroll programs for mainframe computers, programs that process income tax returns and send out welfare checks—programs that today run most of this country. In many ways those programs have become our country. And sometime during those thirty-odd years of being moved from one mainframe computer to another, larger mainframe computer, the original program listings, the source code for thousands of mainframe applications, were just thrown away. We have the object code—the part of the program that machines can read—which is enough to move the software from one type of computer to another. But the source code—the original program listing that people can read, that has details of how these programs actually work—is often long gone, fallen through a paper shredder back in 1967. *There is mainframe software in this country that cost at least $50 billion to develop for which no source code exists today.*

This lack of commented source code would be no big deal if more of those original programmers had expected their programs to outlive them. But hardly any programmer in 1959 expected his payroll application to be still cutting checks in 1999, so nobody thought to teach many of these computer programs what to do when the calendar finally says it's the year 2000. Any program that prints a date on a check or an invoice, and that doesn't have an algorithm for dealing with a change from the twentieth to the twenty-first century, is going to stop working. I know this doesn't sound like a big problem, but it is. *It's a very big problem.*

Looking for a growth industry in which to invest? Between now and the end of the decade, every large company in America either will have to find a way to update its mainframe software or will have to write new software from scratch. New firms will appear dedicated to the digital archaeology needed to update old software. Smart corporations will trash their old software altogether and start over. Either solution is going to cost lots more than it did to write the software in the first place. And all this new mainframe software will have one thing in common: it won't run on a mainframe. Mainframe computers are artifacts of the 1960s and 1970s. They are kept around mainly to run old software and to gladden the hearts of MIS directors who like to think of themselves as mainframe gods. Get rid of the old software, and there is no good reason to own a mainframe computer. The new software will run faster, more reliably, and at one-tenth the cost on a desktop workstation, which is why the old IBM is doomed.

"But workstations will never run as reliably as mainframes," argue the old-line corporate computer types, who don't know what they are talking about. Workstations today can have as much computing power and as much data storage as mainframes. Ten years from now, they'll have even more. And by storing copies of the same corporate data on duplicated machines in separate cities or countries and connecting them by high-speed networks, banks, airlines, and all the other other big transaction processors that still think they'd die without their mainframe computers will find their data are safer than they are now, trapped inside one or several mainframes, sitting in the same refrigerated room in Tulsa, Oklahoma.

Mainframes are old news, and the $40 billion that IBM brings in each year for selling, leasing, and servicing mainframes will be old news too by the end of the decade.

There is going to be a new IBM, I suppose, but it probably won't be the company we think of today. The new IBM *should* be a quarter the size of the current model, but I doubt that current management has the guts to make those cuts in time. The new IBM is already at a disadvantage, and it may not survive, with or without Bill Gates.

So much for mainframes. What about personal computers? PCs, at least as we know them today, are doomed too. That's because the chips are coming.

While you and I were investing decades alternately destroying brain cells and then regretting their loss, Moore's Law was enforcing itself up and down Silicon Valley, relentlessly demanding that the number of transistors on a piece of silicon double every eighteen months, while the price stayed the same. Thirty-five years of doubling and redoubling, thrown together with what the lady at the bank described to me as "the miracle of compound interest," means that semiconductor performance gains are starting to take off. Get ready for yet another paradigm shift in computing.

Intel's current top-of-the-line 80486 processor has 1.2 million transistors, and the 80586, coming in 1992, will have 3 million transistors. Moore's Law has never let us down, and my sources in the chip business can think of no technical reason why it should be repealed before the end of the decade, so that means we can expect to see processors with the equivalent of 96 *million* transistors by the year 2000. Alternately, we'll be able to buy a dowdy old 80486 processor for $11.

No single processor that can be imagined today needs 96 million transistors. The reality of the millennium processor is that it will be a lot smaller than the processors of today, and smaller means faster, since electrical signals don't have to travel as far inside the chip. In keeping with the semiconductor makers' need to add value continually to keep the unit price constant, lots of extra circuits will be included in the millennium processor—circuits that have previously been on separate plug-in cards. Floppy disk controllers, hard disk controllers, Ethernet adapters, and video adapters are already leaving their separate circuit cards and moving as individual chips onto PC motherboards. Soon they will leave the motherboard and move directly into the microprocessor chip itself.

Hard disk drives will be replaced by memory chips, and then those chips too will be incorporated in the processor. And there will still be space and transistors left over—space enough eventually to gang dozens of processors together on a single chip.

Apple's Macintosh, which used to have more than seventy separate computer chips, is now down to fewer than thirty. In two years, a Macintosh will have seven chips. Two years after that, the Mac will be two chips, and Apple won't be a computer company anymore. By then Apple will be a software company that sells operating systems and applications for single-chip computers made by Motorola. The MacMotorola chips themselves may be installed in desktops, in notebooks, in television sets, in cars, in the wiring of houses, even in wristwatches. Getting the PC out of its box will fuel the next stage of growth in computing. Your 1998 Macintosh may be built by Nissan and parked in the driveway, or maybe it will be a Swatch.

Forget about keyboards and mice and video displays, too, for the smallest computers, because they'll talk to you. Real-time, speaker-independent voice recognition takes a processor that can perform 100 million computer instructions per second. That kind of performance, which was impossible at any cost in 1980, will be on your desktop in 1992 and on your wrist in 1999, when the hardware will cost $625. That's for the Casio version; the Rolex will cost considerably more.

That's the good news. The bad news comes for companies that today build PC clones. When the chip literally becomes the computer, there will be no role left for computer manufacturers who by then would be slapping a chip or two inside a box with a battery and a couple of connectors. Today's hardware companies will be squeezed out long before then, unable to compete with the economics of scale enjoyed by the semiconductor makers. Microcomputer companies will survive only by becoming resellers, which means accepting lower profit margins and lower expectations, or by going into the software business.

On Thursday night, April 12, 1991, eight top technical people from IBM had a secret meeting in Cupertino, California, with John Sculley, chairman of Apple Computer. Sculley showed them an IBM PS/2 Model 70 computer running what appeared to be Apple's System 7.0 software. What the computer was actually running was yet another Apple operating system code-named Pink, intended to be run on a number of different types of microprocessors. The eight techies were there to help decide whether to hitch IBM's future to Apple's software.

Sculley explained to the IBMers that he had realized Apple could never succeed as a hardware company. Following the model of Novell, the network operating system company, Apple would have to live or die by its software. And living, to a software company, means getting as many hardware companies as possible to use your operating system. IBM is a *very big* hardware company.

Pink wasn't really finished yet, so the demo was crude, the software was slow, the graphics were especially bad, but it worked. The IBM experts reported back to Boca Raton that Apple was onto something.

The talks with Apple resumed several weeks later, taking place sometimes on the East Coast and sometimes on the West. Even the Apple negotiators scooted around the country on IBM jets and registered in hotels under assumed names so the talks could remain completely secret.

Pink turned out to be more than an operating system. It was also an object-oriented development environment that had been in the works at Apple for three years, staffed with a hundred programmers. Object orientation was a concept invented in Norway but perfected at Xerox PARC to allow large programs to be built as chunks of code called objects that could be mixed and matched to create many different types of applications. Pink would allow the same objects to be used on a PC or a mainframe, creating programs that could be scaled up or down as needed. Combining objects would take no time at all either, allowing applications to be written faster than ever. Writing Pink programs could be as easy as using a mouse to move object icons around on a video screen and then linking them together with lines and arrows.

IBM had already started its own project in partnership with Metaphor Computer Systems to create an object-oriented development environment called Patriot. Patriot, which was barely begun when Apple revealed the existence of Pink to IBM, was expected to take 500 man-years to write. What IBM would be buying in Pink, then, was a 300 man-year head start.

In late June, the two sides reached an impasse, and talks broke down. Jim Cannavino, head of IBM's PC operation, reported to IBM chairman John Akers that Apple was asking for too many concessions. "Get back in there, and do whatever it takes to make a deal," Akers ordered, sounding unlike any previous chairman of IBM. Akers knew that the long-term survival of IBM was at stake.

On July 3, the two companies signed a letter of intent to form a jointly owned software company that would continue development of Pink for computers of all sizes. To make the deal appear as if it went two ways, Apple also agreed to license the RISC processor from IBM's RS/6000 workstation, which would be shrunk from five chips down to two by Motorola, Apple's longtime supplier of microprocessors. Within three years, Apple and IBM would be building computers using the same processor and running the same software—software that would look like Apple's Macintosh, without even a hint of IBM's Common User Access interface or its Systems Application Architecture programming guidelines. Those sacred standards of IBM were effectively dead because Apple rightly refused to be bound by them. Even IBM had come to realize that market share makes standards; companies don't. The only way to succeed in the future will be by working seamlessly with all types of computers, even if they are made by competitors.

This deal with Apple wasn't the first time that IBM had tried to make a quantum leap in system software. In 1988, Akers had met Steve Jobs at a birthday party for Katherine Graham, owner of *Newsweek* and the *Washington Post*. Jobs took a chance and offered Akers a demo of NeXTStep, the object-oriented interface development system used in his NeXT Computer System. Blown away by the demo, Akers cut the deal with NeXT himself and paid $10 million for a NeXTStep license.

Nothing ever came of NeXTStep at IBM because it could produce only graphical user interfaces, not entire applications, and because the programmers at IBM couldn't figure how to fit it into their raison d'être—SAA. But even more important, the technical people of IBM were offended that Akers had imposed outside technology on them from above. They resented NeXTStep and made little effort to use it. Bill Gates, too, had argued against NeXTStep because it threatened Microsoft. (When *InfoWorld*'s Peggy Watt asked Gates if Microsoft would develop applications for the NeXT computer, he said, "Develop for it? I'll piss on it.")

Alas, I'm not giving very good odds that Steve Jobs will be the leader of the next generation of personal computing.

The Pink deal was different for IBM, though, in part because NeXTStep had failed and the technical people at IBM realized they'd thrown away a three-year head start. By 1991, too, IBM was a battered company, suffering from depressed earnings and looking at its first decline in sales since 1946. A string of home-grown software fiascos had IBM so unsure of what direction to move in that the company had sunk to licensing nearly every type of software and literally throwing it at customers, who could mix and match as they liked. "Want an imaging model? Well, we've got PostScript, GPI, and X-Windows—take your pick." Microsoft and Bill Gates were out of the picture, too, and IBM was desperate for new software partnerships.

IBM has 33,000 programmers on its payroll but is so far from leading the software business (and knows it) that it is betting the company on the work of 100 Apple programmers wearing T-shirts in Mountain View, California.

Apple and IBM, caught between the end of the mainframe and the ultimate victory of the semiconductor makers, had little choice but to work together. Apple would become a software company, while IBM would become a software and high-performance semiconductor company. Neither company was willing to risk on its own the full cost of bringing to market the next-generation computing environment ($5 billion, according to Cringely's

Second Law). Besides, there weren't any other available allies, since nearly every other computer company of note had already joined either the ACE or SPARC alliances that were Apple and IBM's competitors for domination of future computing.

ACE, the Advanced Computing Environment consortium, is Microsoft's effort to control the future of computing and Compaq's effort to have a future in computing. Like Apple-IBM, ACE is a hardware-software development project based on linking Microsoft's NT (New Technology) operating system to a RISC processor, primarily the R-4000, from MIPS Computer Systems. In fact, ACE was invented as a response to IBM's Patriot project before Apple became involved with IBM.

ACE has the usual bunch of thirty to forty Microsoft licensees signed up, though only time will tell how many of these companies will actually offer products that work with the MIPS/Microsoft combination.

But remember that there is only room for two standards; one of these efforts is bound to fail.

∽∽

In early 1970, my brother and I were reluctant participants in the first draft lottery. I was hitchhiking in Europe at the time and can remember checking nearly every day in the *International Herald Tribune* for word of whether I was going to Vietnam. I finally had to call home for the news. My brother and I are three years apart in age, but we were in the same lottery because it was the first one, meant to make Richard Nixon look like an okay guy. For that year only, every man from 18 to 26 years old had his birthday thrown in the same hopper. The next year, and every year after, only the 18-year-olds would have their numbers chosen. My number was 308. My brother's number was 6.

Something very similar to what happened to my brother and me with the draft also happened to nearly everyone in the personal computer business during the late 1970s. Then, there were thousands of engineers and programmers and would-be entrepreneurs who had just been waiting for something like the personal computer to come along. They quit their jobs, quit their schools, and started new hardware and software companies all over the place. Their exuberance, sheer numbers, and willingness to die in human wave technology attacks built the PC business, making it what it is today.

But today, everyone who wants to be in the PC business is *already in it*. Except for a new batch of kids who appear out of school each year, the only new blood in this business is due to immigration. And the old blood is getting tired—tired of failing in some cases or just tired of working so hard and now ready to enjoy life. The business is slowing down, and this loss of energy is the greatest threat to our computing future as a nation. Forget about the Japanese; their threat is nothing compared to this loss of intellectual vigor.

Look at Ken Okin. Ken Okin is a *great* hardware engineer. He worked at DEC for five years, at Apple for four years, and has been at Sun for the last five years. Ken Okin is the best-qualified computer hardware designer in the world, but Ken Okin is typical of his generation. Ken Okin is tired.

"I can remember working fifteen years ago at DEC," Okin said. "I was just out of school, it was 1:00 in the morning, and there we were, testing the hardware with all these logic analyzers and scopes, having a ball. 'Can you believe they are paying for us to play?' we asked each other. Now it's different. If I were vested now, I don't know if I would go or stay. But I'm not vested—that will take another four years—and I want my *fuck you money*."

Staying in this business for fuck you money is staying for the wrong reason.

Soon, all that is going to remain of the American computer industry will be high-performance semiconductors and software, but I've just predicted that we won't even have the energy to stay ahead in software. Bummer. I guess this means it's finally my turn to add some value and come up with a way out of this impending mess.

The answer is an increase in efficiency. The era of start-ups built this business, but we don't have the excess manpower or brainpower anymore to allow nineteen out of twenty companies to fail. We have to find a new business model that will provide the same level of reward without the old level of risk, a model that can produce blockbuster new applications without having to create hundreds or thousands of tiny technical bureaucracies run by unhappy and clumsy administrators as we have now. We have to find a model that will allow entrepreneurs to cash out without having to take their companies public and pretend that they ever meant more than working hard for five years and then retiring. We started out, years ago, with Dan Fylstra's adaptation of the author-publisher model, but that is not a flexible or rich enough model to support the complex software projects of the next decade. Fortunately, there is already a business model that has been perfected and fine-tuned over the past seventy years, a business model that will serve us just fine. Welcome to Hollywood.

The world eats dinner to U.S. television. The world watches U.S. movies. It's all just software, and what works in Hollywood will work in Silicon Valley too. Call it the *software studio*.

Today's major software companies are like movie studios of the 1930s. They finance, produce, and distribute their own products. Unfortunately, it's hard to do all those things well, which is why Microsoft reminds me of Disney from around the time of *The Love Bug*. But the movie studio of the 1990s is different; it is just a place where directors, producers, and talent come and go—only the infrastructure stays. In the computer business, too, we've held to the idea that every product is going to live forever. We should be like the movies and only do sequels of hits. And you don't have to keep the original team together to do a sequel. All you have to do is make sure that the new version can read all the old product files and that it feels familiar.

The software studio acknowledges that these start-up guys don't really want to have to create a large organization. What happens is that they reinvent the wheel and end up functioning in roles they think they are supposed to like, but most of them really don't. And because they are performing these roles—pretending to be CEOs—they aren't getting any programming done. Instead, let's follow a movie studio model, where there is central finance, administration, manufacturing, and distribution, but nearly everything else is done under contract. Nearly everyone—the authors, the directors, the producers—works under contract. And most of them take a piece of the action and a small advance.

There are many advantages to the software studio. Like a movie studio, there are established relationships with certain crafts. This makes it very easy to get a contract programmer, writer, marketer, etc. Not all smart people work at Apple or Sun or Microsoft. In fact, most smart people don't work at *any* of those companies. The software studio would allow program managers to find the very best person for a particular job. A lot of the scrounging is eliminated. The programmers can program. The would-be moguls can either start a studio of their own or package ideas and talent together just like independent movie producers do today. They can become minimoguls and make a lot of money, but be responsible for at most a few dozen people. They can be Steven Spielberg or George Lucas to Microsoft's MGM or Lotus's Paramount.

We're facing a paradigm shift in computing, which can be viewed either as a catastrophe or an opportunity. Mainframes are due to die, and PCs and workstations are colliding. Processing power is about to go off the scale, though we don't seem to know what to do with it. The hardware business is about to go to hell, and the people who made all this possible are fading in the stretch.

What a wonderful time to make money!

Here's my prescription for future computing happiness. The United States is losing ground in nearly every area of computer technology except software and microprocessors. And guess what? About the only computer technologies that are likely to show substantial growth in the next decade are—software and microprocessors! The rest of the computer industry is destined to shrink.

Japan has no advantage in software, and nothing short of a total change of national character on their part is going to change that significantly. One really remarkable thing about Japan is the achievement of its craftsmen, who are really artists, trying to produce perfect goods without concern for time or expense. This effect shows, too, in many large-scale Japanese computer programming projects, like their work on fifth-generation knowledge processing. The team becomes so involved in the grandeur of their concept that they never finish the program. That's why Japanese companies buy American movie studios: they can't build competitive operations of their own. And Americans sell their movie studios because the real wealth stays right here, with the creative people who invent the software.

The hardware business is dying. Let it. The Japanese and Koreans are so eager to take over the PC hardware business that they are literally trying to buy the future. But they're only buying the past.

BUT WAIT, THERE'S MORE!

January 1996: Five years have passed since I finished writing this book, and much to my relief the world of computing continues pretty much along the course I've already described. The trend toward graphical desktop computing and away from the mainframe is, if anything, accelerating as America and the world finally realizes the vision of Xerox PARC, circa 1973. Old-line makers of big computers—companies like IBM, DEC, Prime, and Wang—are desperately reinventing themselves or dead. Makers of me-too PC clones are failing by the dozen as commoditization and competition drives their profit margins toward zero. The Japanese and Koreans still aren't making any money selling PCs and are in fact coming under siege in their own countries by leaner and far more efficient American manufacturers like Compaq and Dell. And as predicted, the companies that are doing really well are those that make software, advanced semiconductors, or leading-edge hardware.

But a few things *have* changed in five years, most of them for the better. Bill Gates of Microsoft has finally opened his wallet, giving more than $18 million to charity (I take personal credit for that one). Bill married, too, defying my prediction that he would stay single. But the wedding to Microsoft product manager Melinda French, which took place on January 1, 1994, at first appeared to be more merger than marriage, since it came mainly in response to a demand by Mary Gates that she see her son married in her lifetime (she died of cancer five months after the wedding). There may be room after all for true love in a digital world, however, because 1996 saw Chairman Bill finally become a father, though he was back at work that afternoon.

Five years ago I was among the first to predict IBM's fall and Microsoft's subsequent rise, both of which came to pass. We're in the Age of Microsoft and the 17,000-plus minions of Bill Gates rule the world of personal computing more completely than IBM ever did. Eighty-five percent of the computers in the world run Microsoft operating systems. More than half of the money spent on computer applications goes directly to Microsoft, which is by far the largest and most profitable software company in the world.

The Microsoft culture has changed a bit as the company and its founders age. Bill Gates has gone from gang leader to father figure to godhead. He's rarely seen around the Microsoft campus these days, not because he isn't working there, but just because the place is so darned big. Bill's Lexus is still parked every day in Microsoft's only assigned parking space—assigned not so much because Bill is supposed to be better than you or me, but because Bill is *richer* than anyone. The richest man in the world has four television cameras trained on his car just in case that's where he is abducted by kidnappers. And it could happen: the 1993 abduction of Chuck Geschke from the Adobe Systems' parking lot brought a new sense of reality to high-tech honchos everywhere.

The Age of Microsoft dates, I believe, from a moment in 1989 when executive vice-president Steve Ballmer borrowed some money. Prior to that moment, Microsoft had all the elements necessary for global digital dominance except the will to make it happen. Ballmer's mortgage signified that there was finally a will to go with the way.

Ballmer was Bill Gates's Harvard buddy and Microsoft's twentieth employee. When Ballmer joined Microsoft in those early days, he didn't even have an office, but was granted space at one end of the sofa near Bill's desk. Ballmer, a former brand manager at Procter & Gamble, represented Microsoft's new business orientation, an orientation that surged after 1989. By that year, Ballmer was running Microsoft's operating system business and, just as Bill had, he came to the blinding realization that IBM no longer controlled the PC business—Microsoft did. At that moment of clarity, it was not hard to look ahead and see that careful control of the operating system business could yield enormous profits. Microsoft, which was then a $1 billion company, was poised for incredible growth—although only a few people knew it at the time.

So Ballmer took a chance. He borrowed everything he could against his Microsoft stock, stock options, and his every other possession. In all, Ballmer was able to borrow $50 million and he used every cent to buy more Microsoft shares. This is radical behavior for a PC executive. Most of these folks are continually engaged in selling their company stock, not buying it. Bill Gates, for example, sells 1 million Microsoft shares per quarter, yielding an average of $300 million per year for outside investing, building houses, buying works of art—you know, the usual. High-tech moguls like Gates are usually concerned with the orderly diversification of their wealth; in contrast, Ballmer was betting his entire fortune on Microsoft. This is the only instance I can recall of such behavior.

There's something about betting every penny you have in the world that helps with focus, and Microsoft has been very focused during the 1990s. As a result, Steve Ballmer is now Microsoft's third billionaire, joining Bill Gates and Paul Allen. His shares have increased in value by twenty times since 1989.

Alas, there are hardly any other software billionaires these days and that, too, can be traced to Microsoft. Fearing the growing power of Microsoft, the PC software industry has been madly consolidating and restructuring, trying to find through alliances and mergers some way to respond to what Pete Peterson not long ago described to me as "the menace of Microsoft."

Pete Peterson was until mid-1992 head of nearly everything at WordPerfect Corp., makers of what was then the most popular word processing software for PCs. Peterson and his wife and many, many children still live right next door to WordPerfect headquarters in Orem, Utah. From his living room, Pete can see straight into his old office. WordPerfect was late in making its jump to writing Windows software, preferring instead to stick with its immensely profitable MS-DOS word processor. By the time Peterson and his company began to pay real attention to Windows, Microsoft already had a Windows word processor on the market and Windows applications had suddenly turned into 30 percent of the world processing business.

WordPerfect for Windows finally appeared in early 1992, at a time when Microsoft was already shipping the second generation of its Word for Windows—clearly (and for the first time ever) a superior product to WordPerfect. Technically outgunned, WordPerfect founders Alan Ashton and Bruce Bastian, who each owned 49 percent of the company, proposed to spend some of the more than $100 million they had in the bank to try outmarketing Microsoft. Peterson, who owned only 1 percent of WordPerfect, wanted to put the money into further product development, arguing that not even WordPerfect could outmarket a company that had both better technology and all the money in the world with which to promote it. Peterson was right, but he lost the fight. Ashton and Bastian fumbled their attempt at running the company and were soon so intimidated by Microsoft that they sold WordPerfect to Novell in late 1993 for $850 million.

What we are seeing is just the start of a process of maturation that will take several more years to finish. It parallels what happened in the American automobile industry earlier in this century. In 1920 there were about 300 American companies building automobiles. By 1930, this number had dropped to 25. By 1940 there were 10. Today there are 3. The same thing is happening in the software business, only faster. We took seventy years to accelerate from zero to 100 million cars, but we torqued to 100 million PCs in less than twenty years. So expect the software shakeout to take five years, tops, starting back in 1993.

The shakeout in the American automobile business began when Alfred Sloan brought together a dozen car companies under one name—General Motors—creating a company so big, with such economies of scale, that the other car companies had to grow too, or else be unable to compete. In the software business of today, Microsoft is General Motors, with nearly half the industry sales. Novell is Ford. Lotus Development used to be Chrysler until they sold out to IBM in 1995. Everyone else is either a merger candidate, a maker of expensive custom cars, or doomed.

In 1993 all the smaller companies wanted to merge with Novell, because Novell was considered the only company that could stand up to Microsoft. Even Lotus wanted to merge with Novell, though Lotus CEO Jim Manzi's ego demanded that we call it Lotus buying Novell, rather than the other way around.

Alas, after spending $850 million for WordPerfect and $150 million for Borland's Quattro Pro spreadsheet, Novell chairman Ray Noorda seemed to lose all his smarts. Novell embraced Windows, sure, but in doing so announced that it would no longer upgrade its DOS word processor, throwing millions of customers into the arms of Microsoft. Two years of losses later, Novell put

WordPerfect up for sale again. Corel bought the dregs for $150 million.

Mistakes were made by many Microsoft competitors, not just WordPerfect. Take Lotus Development, where Manzi didn't like Gates, didn't want Windows to succeed, and thought that having his company ignore the Windows market would simply make that market go away. It didn't. By the time 1-2-3 for Windows hit the market, it was three generations behind Microsoft's Excel spreadsheet and hard put to catch up.

There is a difference, though, between WordPerfect and Lotus. Although both companies were late getting into the Windows market, WordPerfect (now Corel) has a much greater likelihood of catching up with Microsoft again, because those nice Mormon kids who hike down the road from Brigham Young University each year after they graduate are used to doing as they are told. At Lotus, *not* doing as you are told seems to be rewarded; that will continue to hurt the company.

According to programmers still on the job at Lotus, if you meet your deadlines for producing new applications, nobody notices. The way to generate five-figure bonuses and executive adoration is not to meet deadlines, but to miss them. Once the situation looks truly hopeless, a few weeks of all-nighters, appearing in the office sloppy and unshaven, generally produces a product that is late, often missing a few features, but is gratefully received. Company reorganizations and layoffs, which always slow product development, are seen at Lotus as a tool for bonus generation, not a means of saving money or increasing efficiency for the company.

It's a corporate culture like this that explains how Manzi could take control of Lotus in 1985 at a time when the company was larger than Microsoft, and ten years later have the company be one quarter Microsoft's size. In that time Lotus's share price doubled while Microsoft's rose fortyfold. Even in a rising industry, Manzi was a failure, which explains why IBM was willing to take a chance on a hostile takeover of Lotus in July 1995.

There's an adage in the computer business that the hostile takeover of a software company is impossible. What is a software company, after all, but a rented office filled with nerds? Attack the company and the nerds will flee, taking with them the real corporate assets—their programming skills. So why would an old-line company like IBM—a company that had never mounted a hostile takeover in its seventy-year history and certainly wasn't known for radical thinking of any sort—mount just such an attack on Lotus Development Corp., offering $3.3 billion for the spreadsheet pioneer?

This kind of takeover had simply never happened before. In making its tender offer for Lotus, IBM was treading new ground and, from a traditional perspective, it was taking a terrible risk. At least that's the way things looked.

The fact that staid old IBM was mounting the first such takeover was especially odd. Oracle Systems might make such a move, or maybe even Microsoft, but IBM? This strange move suggested that there is more here than might be expected, and in fact it came down to semantics. There was a takeover in progress, that's for sure, but IBM was counting on the technical geniuses at Lotus viewing it as a friendly takeover. The only hostility IBM wanted to express was toward CEO Manzi, who opposed the deal.

The personnel of software companies can be generally divided into two groups—techies and suits. Manzi, an M.B.A. who came to Lotus originally as a consultant from McKinsey & Co., was a suit among suits, proud in the past to the point of arrogance about how he didn't understand the inner workings of his company's products—and about how he didn't have to understand those workings. He had techies for that. This is not the sort of attitude to endear the boss to his programmers, a group that generally see suits as a necessary evil kept around mainly to count the money.

There was no doubt Lotus was for sale: Manzi was widely known to have offered the company to AT&T in 1994 for $100 per share. So the issue was not management continuity, but price. Manzi, who was also Lotus's largest shareholder, wanted more than the $60 per share being offered by IBM.

Even at $60 per share, this would have been the biggest software merger ever, but what IBM wanted was neither Lotus 1-2-3, the venerable PC spreadsheet program, nor any of Lotus's other PC applications. IBM wanted a Lotus product called Notes, the leading example of a new software category known as groupware—software that allows groups of workers to communicate with one another and access the same data. Lotus and IBM were both confident that groupware would be central to businesses in the next decade.

Ironically, IBM had once owned a piece of Notes. Unsure of Notes' success, Manzi sold to IBM in 1991 the right to a third of all Notes future revenue for $40 million. A couple of years later, when Manzi realized how strategic Notes was to Lotus's future, he retrieved these rights from IBM by repaying the money and promising to produce applications for IBM's OS/2 operating system.

Now IBM wanted back what it once had, along with the other two thirds. The price increase from $40 million to $3.3 billion just reflected how much the PC software market had changed in four years and how vital to its future success IBM saw Notes.

IBM was not the only company attracted to Notes. AT&T liked Notes, too. The only big player in the software business guaranteed not to be interested in Lotus was Microsoft, which would face antitrust problems from its head-to-head competition with Lotus in the spreadsheet, word processing, and database categories. But in fact, Microsoft had already had its chance to buy Notes: Manzi offered the whole product to Microsoft in 1989 for $20 million, but Microsoft was then willing only to pay $15 million.

The result of all this wrangling is that IBM—even an old and tired IBM—still tends to get what it wants. Manzi changed his mind in exchange for a slightly higher price per share and a golden parachute deal. He left IBM four months later with $87 million, which is a lot of money, but perhaps a tenth of what he'd have had if Lotus had been sold to Microsoft back in 1985, as detailed in chapter 8.

As for IBM, it is easy to forget, given all the bad press the company has received in the last few years, that IBM is still the biggest computer company in the world. What IBM isn't, anymore, is the biggest *personal* computer company. IBM still sells around $9 billion worth of PCs and peripherals each year, but its role as a leader in the PC industry has become inconsequential. Compaq, Apple, and Packard Bell are bigger. While the people of IBM slowly realized that something has changed, their company actually lost its controlling position in the industry back in 1987. IBM is so big and so slow to come to any understanding—much like a brontosaurus that needs an extra brain at the base of its tail just to keep those nerve impulses flowing—that it took almost six years for the truth to sink in.

Bill Gates saw the failure of OS/2 1.0 in 1987 as his chance to take the technical leadership of the PC industry away from IBM. He accomplished this in the late 1980s by introducing successive versions of Windows, each better than the one before. Gates replaced a hardware standard with a software standard, which sounds odd but is actually the way these things are nearly always done.

The trend in information technologies is to first solve a problem with expensive, dedicated hardware, then with general-purpose nondedicated hardware, and finally with software. The first digital computers, after all, weren't really computers at all: they were custom-built machines for calculating artillery trajectories or simulating atomic bombs. The fact that they used digital circuits was almost immaterial, since the early machines could not be easily programmed. The next generation of computers still relied on custom hardware, but could be programmed for many types of jobs. Today's computers often substitute software, in the form of emulators, for what was originally done in custom hardware.

That's what Microsoft has done to the PC business. It doesn't matter anymore whether you have a PC, ISA, EISA, PCI, or Micro Channel bus, because the software looks the same. The real work is not accomplished, after all, by the type of computer, video card, or floppy drive you have, but by the software. And that software is generally some version of Microsoft Windows. PC users now buy more Windows applications than MS-DOS applications, so Windows is now the clear standard, a switch that Microsoft made final by dropping DOS altogether as a separate product when Windows 95 shipped. Oh, there's a version of MS-DOS (DOS 7.0) lurking inside Windows 95, but Microsoft pretends there isn't.

IBM didn't learn the lesson that it no longer set the PC standards until years had passed and its market share had eroded from more than 25 percent to less than 10 percent. Over four years the company posted losses of $20 billion and its market capitalization dropped by another $30 billion. The company also changed its leadership, changed its culture, and cut itself in half, but not without a lot of kicking and screaming along the way.

In the fall of 1993, I was personally involved for a moment in the decline of IBM. At that time, the company was reeling from declines in mainframe sales and total failure in both the PC and workstation markets. Chairman John Akers decided to convene a high-level powwow of IBM's best brains from around the world. They'd meet in a secret retreat and calculate a course out of the current dilemma. And to set the tone for the meeting without actually inviting any outsiders to join in, Akers commissioned a special video production: a TV crew was sent all over America to poll the best minds about what was wrong with IBM and what could be done about it. Somehow, that TV crew landed on my doorstep in California.

We talked for hours, then I sent them to talk with some of my friends. The TV people seemed very excited about what I had to say. But when they submitted their list of interview subjects to IBM's top managers, the only name struck from the list as unsuitable was mine—Robert X. Cringely. Apparently I had made IBM's enemies list. The exercise was in vain, though, because Akers was fired before the video was ever completed.

Ironically, what ended Akers's career and cost IBM so much money was not, as many people suggest, the responsibility of Microsoft. Sure, IBM lost its advantage to Microsoft in the PC business, but that would never have been enough, by itself, to bring down a CEO. Remember, IBM was an enormous company with many business lines—the PC business never accounted for more than 25 percent of total sales for Big Blue, even in the best years.

The truth was that IBM had doomed itself years before. What almost killed IBM was an enormous accounting error. In the early 1980s, a very clever chief financial officer decided that the way to enhance revenue for IBM's mainframe computer business would be to switch customers from leasing their computers to buying them. As leases ended, customers bought either their mainframes outright or the replacement machine outright, with the result that IBM had an enormous increase in revenue. Sounds great, except IBM finance people made the stupid assumption that these high revenue levels would continue forever. They never anticipated the time when every lease was converted into a purchase. That unanticipated day, which came in the early 1990s, was doomsday for Big Blue.

Still, IBM had plenty of nightmares reacting to the very real menace of Microsoft. Predestined or not, IBM focused intently on the threat posed by Bill Gates. Under Jim Cannavino, who replaced Bill Lowe as head of IBM's PC operation, Big Blue tried to do a Microsoft-like introduction of OS/2 2.0, its next-generation operating system (actually, this was a case of IBM imitating a Microsoft imitation of Apple). Cannavino also tried to sell OS/2 direct to users over the telephone, but no provision was made initially for accepting payment by check or purchase order (an oversight that any company might have made, but then any other company would have quickly fixed the problem; for months, IBM didn't). Cannavino and Akers also came up with a plan over lunch one day to have IBM employees sell OS/2 to their friends and neighbors in a kind of digital Amway operation. Though I asked my 500,000 *InfoWorld* readers about this several times, nobody ever reported buying OS/2 from his or her neighbor.

Covering your ass became the watchword at IBM, where Cannavino was awed by chairman John Sculley's setup at Apple. Sculley had been able to make mistakes for years with impunity, sending a new squadron of subordinates down in flames, instead, at each misstep. Cannavino once asked in an IBM meeting, "How do I get a setup like that?"

Cannavino's imitation of Apple was called the IBM Personal Computer Company. In mid-1992, IBM effectively spun off its $9 billion personal computer division, giving it greater autonomy to do whatever it would take to compete in the cutthroat PC business, even if that meant taking actions that might hurt other IBM divisions. While Cannavino headed the total operation, he carefully Sculleyfied the new company by placing himself at the top as a kind of holding company chairman, with most of the heat being taken by the new president, a thirty-year IBMer named

Robert Corrigan. Cannavino was carefully insulated from his own plan. And it worked. Or rather, when it didn't work—when the IBM Personal Computer Company didn't meet early goals for production and profitability—it was Corrigan who initially took the fall, not Cannavino. Corrigan survived for only eighteen months; Cannavino lasted thirty. When I interviewed him after his departure from IBM, Cannavino spoke for ninety minutes about his experiences, referring to IBM throughout as "they," never "we." A little bitter, perhaps?

What Cannavino had to be bitter about was watching new IBM chairman Louis Gerstner dismantle the company Cannavino had worked at for twenty-nine years. Worse still, from Cannavino's perspective, Gerstner didn't consider the IBM veteran a suitable heir. After Cannavino's departure, Gerstner replaced him as chief strategist with one of the founders of Boston Chicken, a New England restaurant chain. This sort of hire signified Gerstner's total commitment to change. The old IBM was forever dead. When I visited IBM intergalactic headquarters at Armonk, New York, in the summer of 1995, I couldn't find a single person wearing a suit and tie.

Turmoil was happening at Apple Computer, too. Despite his carefully crafted layers of protection, John Sculley finally lost his job in 1993. After a decade of bad management and not really understanding the technology he was trying to sell, Sculley was finally killed by his only real attempt at technical leadership—the Newton handheld computer. Newton was Sculley's baby and absorbed more than $200 million of Apple's cash before it finally shipped to resoundingly bad reviews early that year.

Newton, which used pen input and handwriting recognition,

didn't do a very good job of recognizing its owner's handwriting. Worse still, there was no compelling Newton application—no program that, all by itself, justified the purchase of a Newton. Of course there was no compelling application at first for the Macintosh either, which also began its life as a spectacular failure. And just like the Mac, Newton probably will find its compelling application and a certain level of success. But John Sculley won't be there to see it. Back in 1985, when the Macintosh was Apple's sales embarrassment, Steve Jobs had taken the fall. But there was no Steve Jobs to be sacrificed in 1993, so Sculley was finally held accountable.

As big companies are wont to do, Apple tried to play Sculley's departure as a positive event, the retirement of an honored executive who wanted new challenges. First, Sculley gave up his CEO position to Mike Spindler, who had done so much to build Apple's European organization. Sculley stayed on a few months as chairman, then later retired from that position too. But some Apple shareholders didn't like the $4 million severance payment Sculley received and filed a lawsuit to regain the money. They also didn't like the fact that Apple bought back Sculley's California house and even purchased the Learjet Sculley used on company business (Sculley owned the plane and Apple had leased it from him). Why did Sculley get such a generous settlement, the shareholders demanded to know?

Sculley got such a good deal because he didn't "retire" at all. Court documents show Sculley was "terminated for cause." He was fired.

Ironically, Sculley was fired at a time when Apple finally seemed to be succeeding in the overall personal computer business. After the fall of Jean-Louis Gassée, Apple moved from a selling a small number of high-priced computers to selling a larger number of lower-priced computers. It was the right move to have made, but the resulting layoff of 1,100 workers showed that, as usual, Apple was winging it. Most other companies that drastically cut their margins would expect to have to lower costs to keep pace, but this need came as a surprise to Apple. They didn't really expect the cheap computers to sell so well.

The bright spot was the introduction of Apple's PowerBook series of notebook computers, which weren't really so low priced and sold 400,000 units ($1 billion!) in their first year on the market. Apple had simply jumped onto another wave—the notebook computer—that was already mature in the MS-DOS world. Once again with the PowerBooks, Apple has managed to get the same customer to buy yet another computer. PowerBooks are not the future—they are part of the past.

Spindler's job, when he took over from Sculley, was to reengineer Apple, turning it from a computer hardware company to a computer software company. As a computer hardware company, Apple faced (and faces) insurmountable odds. One problem was Wall Street, with its inexorable demand for improved earnings per share. Wall Street compares Apple to Compaq, since both companies are of comparable size, and wonders why Apple doesn't make as much profit as Compaq. That's easy: like Compaq, Apple has to design and build computer systems, but unlike Compaq, Apple also has to design and build all its own operating system software. Compaq leaves that part up to Microsoft. So the extra $800 million Apple spends each year on system software hurts their profits in comparison to Compaq, which doesn't have to develop any software. And Apple can't make tons of money on software like Microsoft does, because there is a limit on the number of copies it can sell: Apple can only sell as many copies of its software as there are Macintosh computers and there just aren't that many Macs in use, compared to Windows machines.

Apple would be much better off and certainly more profitable if it could just jettison its hardware business and compete with Microsoft in software. But that's easier said than done, since

Apple's hardware business—factories and all—has hardly any real value. Who wants to buy the right to build and sell computers that can't be sold at a profit? Nobody. Who wants to even buy Apple's state-of-the-art factories? Almost nobody.

A decade ago, if a company like Apple or IBM wanted to sell a factory, potential buyers lined-up, checkbooks in hand. That was because the factories were known for their high quality work. But today we have the International Standards Organization running around the world certifying the quality of factories. Today an ISO 9001 certification means some factory in Malaysia or Mexico is precisely as good as an Apple factory in California or an IBM factory in New York. So these big buildings, which are carried on the company books as having such and such a value, aren't really worth that much. And no CEO wants to take the earnings hit that inevitably comes from throwing the factory away and paying-off all those manufacturing employees. Certainly Spindler didn't want to do it.

Spindler had to go. It's not just that he was paralyzed and unable to make the bold moves needed to save Apple. He had to go just to make it obvious that Apple was doing something to solve its problem. This is a common technique to use on baseball teams. The players are good, yet the team keeps losing, so the owner fires the manager. This gets the attention of the players (in Apple's case, it gets the attention of the workers, customers and, most importantly, the press) and provides motivation. Firing the manager—no matter who the replacement manager is—always wins a few ballgames.

So Spindler was fired, replaced by Gil Amelio, who had done a very good job of turning-around National Semiconductor, a maker of microprocessors and semi-custom integrated circuits. Amelio succeeded at National Semiconductor by getting the company to concentrate on its core businesses, which is exactly what Apple needs to do. So now the Apple workers know that the company is serious about change. They also know that the new leader has no particular friends or political allies among the management groups: every job is equally at risk. The result is that the bad workers are scared away and the good workers are motivated. At least that's the way it is supposed to work.

But not even a new president can quickly change the direction of 17,000 workers. It will take a year for any changes to have real effect on Apple's products. So Amelio's much bigger concern has to be with appearance: how does he make Apple look better to its customers and to the press. This is vitally important because of Apple's perilous inventory situation. At the time I am writing this, Apple has almost $2 billion worth of computers stored in its warehouses. That's $2 billion at today's prices. But computers are always going down in price, so those computers are going down in value. That declining value has to be reflected as a loss on Apple's books. This explains, then, just how bad Apple's 1995 Christmas sales really were.

Normally, Apple would have expected to make 45 percent of its annual profit in the Christmas quarter. They were hoping to make something in excess of $500 million in Christmas 1995, but instead lost $69 million. Worse still, Apple had all those unsold computers in the warehouse. In January the inventory was marked-down by an average of 15 percent as Apple tried to sell it. That 15 percent reduction on a $2 billion inventory means Apple had already accepted another $300 million loss, which means the total hit for the bad Christmas of 1995 was close to $400 million.

But wait, it gets worse! Now consumers are worried about whether Apple will survive. When consumers are worried, they stop buying. Apple had $1.1 billion in cash in January, but more than $300 million of that will now be lost to marked-down inventory. Another mark-down, along with a couple more bad quarters of sales, and Apple is out of cash. That's the fear in Cupertino, and that's why Spindler had to go.

In his last days at Apple, Spindler was trying mightily to sell the company. IBM and Motorola both turned him down. Only Sun Microsystems was at all interested in buying Apple. There were many stories in the American press about these negotiations, explaining why Sun wanted Apple. These stories showed that Apple's sales volume could help Sun get parts cheaper and so make their Unix workstations more price competitive. Having Apple was also supposed to give Sun an advantage in building a $500 network appliance—a stripped-down computer used strictly to communicate on the Internet. All these reasons had some small bearing on Sun's decision to negotiate with Apple, but they weren't the major reason, not even close.

Sometimes business decisions are very personal and this was the case with Sun's interest in Apple. John Doerr is a member of the Sun board of directors. He is also one of the venture capitalists who helped finance Sun in its early days. Doerr is a very smart and legendary venture capitalist. He also wants to run a computer company.

John Doerr has been long critical of Apple's strategy. He has told me many times that he could turn Apple around in six months. If Doerr was willing to tell me that, he was also willing to say the same thing to Scott McNealy, Sun's CEO. And that's where Sun's interest in Apple found its greatest strength. Doerr told McNealy, "Buy Apple, give it to me, and I'll have it running like a watch in six months."

It almost happened, too, but the Sun bid was just too low, so Apple chairman Mike Markkula hired Gil Amelio to do exactly what Doerr would have done: cut costs, focus the company on its core businesses, and finally start an extensive (and profitable) software licensing program.

If Amelio succeeds and Apple survives, it will probably have more to do with the efforts of IBM and Motorola than Amelio. Both companies have now taken Macintosh software licenses that allow them to sublicense the software to other companies. The companies plan to sell all the parts to make a Macintosh, along with the Mac ROMS and system software. This will flood the market with cheap Mac clones by the end of 1997, squeezing Apple out of the hardware business while simultaneously making the company successful in software. At least that's the plan.

But Apple's real future, and the future of the entire PC business, lies in finding new customers—millions of them—through selling whole new types of computers. More about that in the next chapter.

▸ ▸ ▸ ▸ ▸ ▸ ▸ ▸ ▸ ▸ ▸ ▸

CHAPTER SEVENTEEN

DO THE WAVE

We're floating now on surfboards 300 yards north of the big public pier in Santa Cruz, California. As our feet slowly become numb in the cold Pacific water, it's good to ponder the fact that this section of coastline, only fifteen miles from Silicon Valley, is the home of the great white shark and has the highest incidence of shark attacks in the world. Knowing that we're paddling not far from creatures that can bite man and surfboard cleanly in half gives me, at least, a heightened awareness of my surroundings.

We're waiting for a wave, and can see line after line of them approaching from the general direction of Hawaii. The whole ritual of competitive surfing is choosing the wave, riding it while the judges watch, then paddling out to catch another wave. Choose the wrong wave—one that's too big or too small—and you'll either wipe out (fall off the board) or not be able to do enough tricks on that wave to impress the judges. Once you've chosen the wave, there is also the decision of how long to ride it, before heading back out to catch another. Success in competitive surfing comes from riding the biggest waves you can handle, working those waves to the max, but not riding the wave too far—because you get more total points riding a lot of short waves than by riding a few waves all the way to the beach.

Surfing is the perfect metaphor for high-technology business. If you can succeed as a competitive surfer, you can succeed in semiconductors, computers, or biotechnology. People who are astute and technically aware can see waves of technology leaving basic research labs at least a decade before they become commercially viable. There are always lots of technology waves to choose from, though it is not always clear right away which waves are going to be the big ones. Great ideas usually appear years—sometimes decades—before they can become commercial products. It takes that long both to bring the cost of a high-tech product down to where it's affordable by the masses, and it can take even longer before those masses finally perceive a personal or business need for the product. Fortunately for those of us who plan to be the next Steve Jobs or Bill Gates, this means that coming up with the technical soul of our eventual empire is mainly a matter of looking down the food chain of basic research to see what's likely to be the next overnight technosensation a few years from now. The technology is already there; we just have to find it.

Having chosen his or her wave, the high-tech surfer has to ride long enough to see if the wave is really a big one. This generally takes about three years. If it isn't a big wave, if your company is three years old, your product has been on the market for a year, and sales aren't growing like crazy, then you chose either the wrong wave or (by starting to ride the wave too early) the wrong time.

Software has been available on CD-ROM optical disks since the mid-1980s, for example, but that business has become profitable only recently as the number of installed CD-ROM drives has grown into the millions. So getting into the CD-ROM business in 1985 would have been getting on the wave too early.

Steve Jobs has been pouring money into NeXT Inc.—his follow-on to Apple Computer—since 1985 and still hasn't turned a net profit. Steve finally turned NeXT from a hardware business into a software business in 1993 (following the advice in this book), selling NeXTstep, his object-oriented version of the Unix operating system. Unfortunately, Steve didn't give up the hardware business until Canon, his bedazzled Japanese backer, had invested and lost $350 million in the venture. The only reason NeXT survives at all is that Canon is too embarrassed to write off its investment. So, while NeXT survives, Steve Jobs has clearly been riding this particular wave at least five years too long.

All this may need some further explanation if, as I suspect, Jobs attempts an initial public stock offering (IPO) for NeXT in 1996. Riding on the incredible success of its computer-animated feature film *Toy Story*, another Jobs company—Pixar Animation Studios—used an IPO to once again turn Steve into a billionaire. Ironically, Pixar is the company in which Jobs has had the least direct involvement. Pixar's success and a 1996 feeding frenzy for IPOs in general suggest that Jobs will attempt to clean up NeXT's balance sheet before taking the software company public. IPOs are emotional, rather than intellectual, investments, so they play well under the influence of Steve's reality distortion field.

Now back to big business. There are some companies that intentionally stay on the same wave as long as they can because doing so is very profitable. IBM did this in mainframes, DEC did it in minicomputers, Wang did it in dedicated word processors. But look at what has happened to those companies. It is better to get off a wave too early (provided that you have another wave already in sight) than to to ride it too long. If you make a mistake and ride a technology wave too far, then the best thing to do is to sell out to some bigger, slower, dumber company dazzled by your cash flow but unaware that you lack prospects for the future.

Surfing is best done on the front of the wave. That's where the competition is least, the profit margins are highest, and the wave itself provides most of the energy propelling you and your company toward the beach. There are some companies, though, that have been very successful by waiting and watching to see if a wave is going to be big enough to be worth riding; then, by paddling furiously and spending billions of dollars, they somehow overtake the wave and ride it the rest of the way to the beach. This is how IBM entered the minicomputer, PC, and workstation markets. This is how the big Japanese electronics companies have entered nearly every market. These behemoths, believing that they can't afford to take a risk on a small wave, prefer to buy their way in later. But this technique works only if the ride is a long one; in order to get their large investments back, these companies rely on long product cycles. Three years is about the shortest product cycle for making big bucks with this technique (which is why IBM has had trouble making a long-term success of its PC operation, where product cycles are less than eighteen months and getting shorter by the day).

Knowing when to move to the next big wave is by far the most important skill for long-term success in high technology; indeed, it's even more important than choosing the right wave to ride in the first place.

Microsoft has been trying to invent a new style of surfing, characterized by moving on to the next wave but somehow taking the previous wave along with it. Other companies have tried and failed before at this sport (for example, IBM with its Office-Vision debacle, where PCs were rechristened "programmable terminals"). But Microsoft is not just another company. Bill Gates knows that his success is based on the de facto standard of MS-DOS. Microsoft Windows is an adjunct to DOS—it requires that DOS be present for it to work—so Bill used his DOS franchise to popularize a graphical user interface. He jumped to the Windows wave but took the DOS wave along with him. Now he is doing the same thing with network computing, multimedia computing, and even voice recognition, making all these parts adjuncts to Windows, which is itself an adjunct to DOS. Even Microsoft's next-generation 32-bit operating system, Windows NT, carries along the Windows code and emulates MS-DOS.

Microsoft's surfing strategy has a lot of advantages. By building on an installed base, each new version of the operating system automatically sells millions of upgrade copies to existing users and is profitable from its first days on the market. Microsoft's applications also have market advantages stemming from this strategy, since they automatically work with all the new operating system features. This is Bill Gates's genius, what has made him the richest person in America. But eventually even Bill will fail, when the load of carrying so many old waves of innovation along to the next one becomes just too much. Then another company will take over, offering newer technology.

But what wave are we riding right now? Not the wave you might expect. For the world of corporate computing, the transition from the personal computing wave to the next wave, called client-server computing, has already begun. It had to.

⌖

The life cycles of companies often follow the life cycles of their customers, which is no surprise to makers of hair color or disposable diapers, but has only lately occurred to the ever grayer heads running companies like IBM. Most of IBM's customers, the corporate computer folks who took delivery of all those mainframe and minicomputers over the past thirty years, are nearing the end of their own careers. And having spent a full generation learning the hard way how to make cantankerous, highly complex, corporate computer systems run smoothly, this crew-cut and pocket-protectored gang is taking that precious knowledge off to the tennis court with them, where it will soon be lost forever.

Marshall McLuhan said that we convert our obsolete technologies into art forms, but, trust me, nobody is going to make a hobby of collecting $1 million mainframe computers.

This loss of corporate computing wisdom, accelerated by early retirement programs that have become so popular in the leaner, meaner, downsized corporate world of the 1990s, is among the factors forcing on these same companies a complete changeover in the technology of computing. Since the surviving computer professionals are mainly from the 1980s—usually PC people who not only know very little about mainframes, but were typically banned from the mainframe computer room—there is often nobody in the company to accept the keys to that room who really knows what he or she is doing. So times and technologies are being forced to change, and even IBM has seen the light. That light is called client-server computing.

Like every other important computing technology, client-server has taken about twenty years to become an overnight sensation. Client-server is the link between old-fashioned centralized computing in organizations and the newfangled importance of desktop computers. In client-server computing, centralized computers called "servers" hold the data, while desktop computers called "clients" use that data to do real work. Both types of computers are connected by a network. And getting all these dissimilar types of equipment to work together looks to many informed investors like the next big business opportunity in high technology.

In the old days before client-server, the computing paradigm du jour was called "master-slave." The computer—a mainframe or minicomputer—was the master, controlling dozens or even hundreds of slave terminals. The essence of this relationship was that the slaves could do nothing without the permission of the master; turn off that big computer and its many terminals were useless. Then came the PC revolution of the 1980s, in which computers got cheap enough to be bought from petty cash and so everybody got one.

In 1980, virtually all corporate computing power resided in the central computer room. By 1987, 95 percent of corporate computing power lay in desktop PCs, which were increasingly being linked together into local area networks (LANs) to share printers, data, and electronic mail. There was no going back following that kind of redistribution of power, but for all the new-found importance of the PC, the corporate data still lived in mainframes. PC users were generally operating on ad hoc data, copied from last quarter's financial report or that morning's *Wall Street Journal.*

Client-server accepts that the real work of computing will be done in little desktop machines and, for the first time, attempts to get those machines together with real corporate data. Think of it as corporate farming of data. The move to PCs swung the pendulum (in this case, the percentage of available computing power) away from the mainframe and toward the desktop. LANs have helped the pendulum swing back toward the server; they make that fluidity possible. Sure, we are again dependent on centralized data, but that is not a disadvantage. (A century ago we grew our own food too, but do we feel oppressed or liberated by modern farming?) And unlike the old master-slave relationship, PC clients can do plenty of work without even being connected to the mainframe.

Although client-server computing is not without its problems (data security, for one thing, is much harder to maintain than for either mainframes or PCs), it allows users to do things they were never able to do before. The server (or servers—there can be dozens of them on a network, and clients can be connected to more than one server at a time) holds a single set of data that is accessible by the whole company. For the first time ever, *everyone* is using the same data, so they can all have the same information and believe the same lies. It's this turning of data into information (pictures, charts, graphs), of course, that is often the whole point of using computers in business. And the amount of data we are talking about is enormous: far greater than any PC could hold, and *vastly* more than a PC application could sort in reasonable time (that's why the client asked the server—a much more powerful computer—to sort the data and return only the requested information). American Express, for example, has 12 terabytes of data—that's 12,000,000,000,000 bytes—on mainframes, mainframes that they *must* get rid of for financial reasons. So if a company, university, or government agency wants to keep all its data accessible in one place, they need a big server, which is still more often than not a mainframe computer. In the client-server computing business, these old but still useful mainframes are called "legacy systems," because the client-server folks are more or less stuck with them, at least for now.

But mainframes, while good storehouses for centralized data, are not so good for displaying information. People are used to PCs and workstations with graphical users interfaces (GUIs) but mainframes don't have the performance to run GUIs. It's much more cost-effective to use microprocessors, to put part of the application on the desk so the interface is quick. That means client-server.

Client-server has been around for the last ten years, but right now users are changing over in phenomenal numbers. Once you've crossed the threshold, it's a stampede. First it was financial services, then CAD (computer-aided design), now ordinary companies. It's all a matter of managing intellectual inventory more effectively. The nonhierarchical management style—the flat organizational model—that is so popular in the 1990s needs more communication between parts of the company. It requires a corporate information base. You can't do this with mainframes or minicomputers.

The appeal of client-server goes beyond simply massaging corporate data with PC applications. What really appeals to corporate computing honchos is the advent of new types of applications that could never be imagined before. These applications, called groupware, make collaborative work possible.

The lingua franca of client-server computing is called Structured Query Language (SQL), an early-1970s invention from IBM. Client applications running on PCs and workstations talk SQL over a network to back-end databases running on mainframes or on specialized servers. The back-end databases come from companies like IBM, Informix, Oracle, and Sybase, while the front-end applications running on PCs can be everything from programs custom-written for a single corporate customer to general productivity applications like Microsoft Excel or Lotus 1-2-3, spreadsheets to which have been added the ability to access SQL databases.

Middleware is yet another type of software that sits between the client applications and the back-end database. Typically the client program is asking for data using SQL calls that the back-end database doesn't understand. Sometimes middleware makes the underlying mainframe database think it is talking not to a PC or workstation, but to a dumb computer terminal; the middle ware simply acts as a "screen scraper," copying data off the emulated terminal screen and into the client application. Whatever it takes, it's middleware's job to translate between the two systems, preserving the illusion of open computing.

Open computing, also called standards-based computing, is at the heart of client-server. The idea is simple; customers ought to be able to buy computers, networking equipment, and software from whomever offers the best price or the best performance and all those products from all those companies ought to work together right out of the box. When a standard is somewhat open, other people want to participate in the fruits of it. There is a chance for synergy and economies of scale. That's what's happening now in client-server.

What has made America so successful is the development of infrastructure—efficient distribution systems for goods, money, and information. Client-server replicates that infrastructure. Companies that will do well are those that provide components, applications, and services, just as the gas stations, motels, and fast food operations did so well as the interstate highway system was built.

For all its promise, client-server is hard to do. Open systems often aren't as open as their developers would like to think, and the people being asked to write specialized front-end applications inside corporate computing departments often have a long learning curve to climb. Typically, these are programmers who come from a mainframe background and they have terrible trouble designing good, high-performance graphical user interfaces. It can take a year or two to develop the skill set needed to do these kinds of applications, but once the knowledge is there, then they can bang out application after application in short time.

Companies that will do less well because of the migration toward client-server computing are traditional mid-range computer companies, traditional mainframe software companies, publishers of PC personal productivity applications, and, of course, IBM.

◦∾◦

If client-server is the present wave, the next wave is probably extending those same services out of the corporation to a larger audience. For this we need a big network, an Internet. There's that word.

The so-called information superhighway has already gone from being an instrument of oedipal revenge to becoming the high-tech equivalent of the pet rock: everybody's got to have it. Young Senator (later vice-president) Al Gore's information super-highway finally overshadows Old Senator Gore's (Al's father's) interstate highway system of the 1950s; and an idea that not long ago appealed only to Democratic policy wonks now forms the basis of an enormous nonpartisan movement. Republicans and Democrats, movie producers and educators, ad executives and former military contractors, everyone wants their own on-ramp to this digital highway that's best exemplified today by the global Internet. Internet fever is sweeping the world and nobody seems to notice or care that the Internet we're touting has some serious flaws. It's like crossing a telephone company with a twelve-step group. Welcome to the future.

The bricks and mortar of the Internet aren't bricks and mor-tar at all, but ideas. This is probably the first bit of public infra-structure anywhere that has no value, has no function at all, not even a true existence, unless everyone involved is in precise agreement on what they are talking about. Shut the Internet down and there isn't rolling stock, rights-of-way, or railway sta-tions to be disposed of, just electrons. That's because the Internet is really a virtual network constructed out of borrowed pieces of telephone company.

Institutions like corporations and universities buy digital data services from various telephone companies, link those ser-vices together, and declare it to be an Internet. A packet of data (the basic Internet unit of 1200 to 2500 bits that includes source address, destination address, and the actual data) can travel clear around the world in less than a second on this network, but only if all the many network segments are in agreement about what constitutes a packet and what they are obligated to do with one when it appears on their segment.

The Internet is not hierarchical and it is not centrally managed. There is no Internet czar and hardly even an Internet janitor, just a quasi-official body called the Internet Engineering Task Force (IETF), which through consensus somehow comes up with the evolving technical definition of what "Internet" means. Anyone who wants to can attend an IETF meeting, and you can even participate over the Internet itself by digital video. The evolving IETF standards documents, called Requests for Comment (RFCs), are readable on big computers everywhere. There are also a few folks in Virginia charged with handing out to whomever asks for them the dwindling supply of Internet addresses. These address-givers aren't even allowed to say no.

This lack of structure extends to the actual map of the Internet, showing how its several million host computers are connected to each other: such a map simply doesn't exist and nobody even tries to draw one. Rather than being envisioned as a tree or a grid or a ring, the Internet topology is most often described as a cloud. Packets of data enter the cloud in one place, leave it in another, and for whatever voodoo takes place within the cloud to get from here to there no money at all is exchanged. This is no way to run a business.

Exactly. The Internet isn't a business and was never intended to be one. Rather, it's an academic experiment from the 1960s to which we are trying, so far without much success, to apply a business model. But that wasn't enough to stop Netscape Communications, Inc., publishers of software to use data over the Internet's World Wide Web, from having the most successful stock offering in Wall Street history. Thus, less than two years after Netscape opened for business, company founder Jim Clark became Silicon Valley's most recent billionaire.

Today's Internet evolved from the ARPAnet, which was yet

another brainchild of Bob Taylor. As we already know, Taylor was responsible in the 1970s for much of the work at Xerox PARC. In his pre-PARC days, while working for the U.S. Department of Defense, he funded development of the ARPAnet so his researchers could stay in constant touch with one another. And in that Department of Defense spirit of $900 hammers and $2000 toilet seats, it was Taylor who declared that there would be no charges on the ARPAnet for bandwidth or anything else. This was absolutely the correct decision to have made for the sake of computer research, but it's now a problem in the effort of turning the Internet into a business.

Bandwidth flowed like water and it still does. There is no incentive on the Internet to conserve bandwidth (the amount of network resources required to send data over the Internet). In fact, there is an actual disincentive to conserve, based on "pipe envy": every nerd wants the biggest possible pipe connecting him or her to the Internet. (We call this "network boner syndrome"— you know, mine is bigger than yours.) And since the cost per megabit drops dramatically when you upgrade from a 56K leased line (56,000 bits-per-second) to a T-1 (1.544 megabits per second) to a T-3 (45 megabits per second), some organizations deliberately add services they don't really need simply to ratchet up the boner scale and justify getting a bigger pipe.

The newest justification for that great big Internet pipe is called Mbone, the Multimedia Backbone protocol that makes it possible to send digital video and CD-quality audio signals over the Internet. A protocol is a set of rules approved by the IETF defining what various types of data look like on the Internet and how those data are to be handled between network segments. If you participate in an IETF meeting by video, you are using Mbone. So far Mbone is an experimental protocol available on certain network segments, but all that is about to change.

Some people think Mbone is the very future of the Internet,

claiming that this new technology can turn the Internet into a competitor for telephone and cable TV services. This may be true in the long run, five or ten years hence, but for most Internet users today Mbone is such a bandwidth hog that it's more nightmare than dream. That's because the Internet and its fundamental protocol, called TCP/IP (transport control protocol/Internet protocol) operates like a telephone party line and Mbone doesn't practice good phone etiquette.

Just like on an old telephone party line, the Internet has us all talking over the same wire and Mbone, which has to send both sound and video down the line, trying to keep voices and lips in sync all the while, is like that long-winded neighbor who hogs the line. Digital video and audio means a *lot* of data—enough so that a 1.544-megabit-per-second T-1 line can generally handle no more than four simultaneous Mbone sessions (audio-only Mbone sessions like Internet Talk Radio require a little less bandwidth). Think of it this way: T-1 lines are what generally connect major universities with thousands of users each to the Internet, but each Mbone session can take away 25 percent of the bandwidth available for the entire campus. Even the Internet backbone T-3 lines, which carry the data traffic for millions of computers, can handle just over 100 simultaneous real-time Mbone video sessions: hardly a replacement for the phone company.

Mbone video is so bandwidth-intensive that it won't even fit in the capillaries of the Internet, the smaller modem links that connect millions of remote users to the digital world. And while these users can't experience the benefits of Mbone, they share in the net cost, because Mbone is what's known as a dense-mode Internet protocol. Dense mode means that the Internet assumes almost every node on the net is interested in receiving Mbone data and wants that data as quickly as possible. This is assured, according to RFC 1112, by spreading control information (as opposed to actual broadcast content) all over the net, even to nodes that don't want an Mbone broadcast—or don't even have the bandwidth to receive one. This is the equivalent of every user on the Internet receiving a call several times a minute asking if they'd like to receive Mbone data.

These bandwidth concerns will go away as the Internet grows and evolves, but in the short term, it's going to be a problem. People say that the Internet is carrying multimedia today, but then dogs can walk on their hind legs.

There is another way in which the Internet is like a telephone party line: anyone can listen in. Despite the fact that the ARPAnet was developed originally to carry data between defense contractors, there was never any provision made for data security. There simply is no security built into the Internet. Data security, if it can be got at all, has to be added on top of the Internet by users and network administrators. Your mileage may vary.

The Internet is vulnerable in two primary ways. Internet host computers are vulnerable to invasion by unauthorized users or unwanted programs like viruses, and Internet data transmissions are vulnerable to eavesdropping.

Who can read that e-mail containing your company's deepest secrets? Lots of people can. The current Internet addressing scheme for electronic mail specifies a user and a domain server, such as my address (bob@cringely.com). "Bob" is my user name and "cringely.com" is the name of the domain server that accepts messages on my behalf. (Having your own domain [like cringely.com] is considered *very* cool in the Internet world. At last, I'm a member of an elite!) A few years ago, before the Internet was capable of doing its own message routing, the addressing scheme required the listing of all network links between the user and the Internet backbone. I recall being amazed to see back then that an Internet e-mail message from Microsoft to Sun Microsystems at one point passed through a router at Apple Computer, where it could be easily read.

These sort of connections, though now hidden, still exist, and every Internet message or file that passes through an interim router is readable and recordable at that router. It's very easy to write a program called a "sniffer" that records data from or to specific addresses or simply records user addresses and passwords as they go through the system. The writer of the sniffer program doesn't even have to be anywhere near the router. A 1994 break-in to New York's Panix Internet access system, for example, where hundreds of passwords were grabbed, was pulled off by a young hacker connected by phone from out of state.

The way to keep people from reading your Internet mail is to encrypt it, just like a spy would do. This means, of course, that you must also find a way for those who receive your messages to decode them, further complicating network life for everyone. The way to keep those wily hackers from invading Internet domain servers is by building what are called "firewalls"—programs that filter incoming packets, trying to reject those that seem to have evil intent. Either technique can be very effective, but neither is built in to the Internet. We're on our own.

Still, there is much to be excited about the Internet. Many experts are excited about a new Internet programming language from Sun Microsystems called Java. What became the Java language was the invention of a Sun engineer named James Gosling. When Gosling came up with the idea, the language was called Oak, not Java, and it was aimed not at the Internet and the World Wide Web (WWW didn't even exist in 1991) but at the consumer electronics market. Oak was at the heart of *7, a kind of universal intelligent remote control Sun invented but never manufactured. After that it was the heart of an operating system for digital television decoders for Time Warner. This project also never reached manufacturing. By mid-1994, the World Wide Web was big news and Oak became Java. The name change was driven solely by Sun's inability to trademark the name Oak.

Java is what Sun calls an "architecture neutral" language. Think of it this way: If you found a television from thirty years ago and turned it on, you could use it today to watch TV. The picture might be in black and white instead of color, but you'd still be entertained. Television is backward-compatible and therefore architecture neutral. However, if you tried to run Windows 95 on a computer built thirty years ago, it simply wouldn't work. Windows 95 is architecture specific.

Java language applications can execute on many different processors and operating system architectures without the need to rewrite the applications for those systems. Java also has a very sophisticated security model, which is a good thing for any Internet application to have. And it uses multiple program threads, which means you can run more than one task at a time, even on what would normally be single-tasking operating systems.

Most people see Java as simply a way to bring animation to the web, but it is much more than that. Java applets are little applications that are downloaded from a WWW server and run on the client workstation. At this point an applet usually means some simple animation like a clock with moving hands or an interactive map or diagram, but a Java applet can be much more sophisticated than that. Virtually any traditional PC application—like a word processor, spreadsheet, or database—can be written as one or more Java applets. This possibility alone threatens Microsoft's dominance of the software market.

Sun Microsystems' slogan is "the network is the computer," and this is fully reflected in Java, which is very workstation-centric. By this I mean that most of the power lies in the client workstation, not in the server. Java applications can be run from any server that runs the World Wide Web's protocol. This means that a PC or Macintosh is as effective a Java server as any powerful Unix box. And because whole screen images aren't shipped across the network, applets can operate using low-bandwidth dial-up links.

The Internet community is very excited about Java, because it appears to add greater utility to an already existing resource, the World Wide Web. Since the bulk of available computing power lies out on the network, residing in workstations, that's where Java applets run. Java servers can run on almost any platform, which is an amazing thing considering Sun is strictly in the business of building Unix boxes. Any other company might have tried to make Java run only on its own servers—that's certainly what IBM or Apple would have done.

In a year or two, when there are lots of Java browsers and applets in circulation, we'll see a transformation of how the Internet and the World Wide Web are used. Instead of just displaying text and graphical information for us, our web browsers will work with the applets to actually do something with that data. Companies will build mission-critical applications on the web that will have complete data security and will be completely portable and scalable. And most important of all, this is an area of computing that Microsoft does not dominate. As far as I can see, they don't even really understand it yet, leaving room for plenty of other companies to innovate and be successful.

What might be the wave after the Internet is building a real data network that extends multimedia services right into our homes. Since this is an area that is going to require very powerful processor chips, you'd think that Intel would be in the forefront, but it's not. Has someone already missed the wave?

Intel rules the microprocessor business just as Microsoft rules the software business: by being very, very aggressive. Since the American courts have lately ruled in favor of the companies that clone Intel processors, the current strategy of Intel president Andy Grove is to outspend its opponents. Grove wants to speed up product development so that each new family of Intel processors appears before competitors have had a chance to clone the previous family. This is supposed to result in Intel's building the very profitable leading-edge chips, leaving its competitors to slug it out in the market for commodity processors. Let AMD and Cyrix make all the 486 chips they want, as long as Intel is building all the Pentiums.

Andy Grove is using Intel's large cash reserves (the company has more than $2.5 billion in cash and no debt) to violate Moore's Law. Remember Moore's law was divined in the late 1950s by Gordon Moore, now the chairman of Intel and Andy Grove's boss. Moore's Law states that the number of transistors that can be etched on a given piece of silicon will double every eighteen months. This means that microprocessor computing power will naturally double every eighteen months, too. Alternately, Moore's Law can mean that the cost of buying the same level of computing power will be cut in half every eighteen months. This is why personal computer prices are continually being lowered.

Although technical development and natural competition have always fit nicely with the eighteen-month pace of Moore's Law, Andy Grove knows that the only way to keep Intel ahead of the other companies is to force the pace. That's why Intel's P-6 processor appeared in mid-1995, more than a year before tradition dictated it ought to. And the P-7 will appear just two years after that, in 1997.

This accelerated pace is accomplished by running several development groups in parallel, which is incredibly expensive. But the same idea of spending, spending, spending on product development is what President Reagan used to force the end of communism by simply spending faster than his enemies could on new weapons. Any Grove figures that what worked for Reagan will also work for Intel.

But what if the computing market takes a sudden change of direction? If that happens, wouldn't Intel still be ahead of its competitors, but ahead of them at running in the *wrong direction?* That's exactly what I think is happening right now.

Intel's strategy is based on the underlying idea that the computing we'll do in the future is a lot like the computing we've done in the past. Intel sees computers as individual devices on desktops, with local data storage and generally used for stand-alone operation. The P-6 and P-7 computers will just be more powerful versions of the P-5 (Pentium) computers of today. This is not a bad bet on Intel's part, but sometimes technology does jump in a different direction, and that's just what is happening right now, with all this talk of a digital convergence of computing and communication.

The communications world is experiencing an explosion of bandwidth. Fiber-optic digital networks and new technologies like asynchronous transfer mode (ATM) networking is leading us toward a future where we'll mix voice, data, video, and music, all on the same lines that today deliver either analog voice or video signals. The world's telephone companies are getting ready to offer us any type of digital signal we want in our homes and businesses. They want to deliver to us everything from movies to real-time stock market information, all through a box that in America is being called a "set-top device."

What's inside this set-top device? Well, there is a microprocessor to decode and decompress the digital data-stream, some memory, a graphics chip to drive a high-resolution color display, and system software. Sounds a lot like a personal computer, doesn't it? It sure sounds that way to Microsoft, which is spending millions to make sure that these set-top devices run Windows software.

With telephone and cable television companies planning to

roll out these new services over the next two to three years, Intel has not persuaded a single maker of set-top devices to use Intel processors. Instead, Apple, General Instrument, IBM, Scientific Atlanta, Sony, and many other manufacturers have settled on Motorola's PowerPC processor family. And for good reason, because a PowerPC 602 processor costs only $20 each in large quantities, compared with more than $80 for a Pentium.

Intel has been concentrating so hard on the performance side of Moore's Law that the company has lost sight of the cost side. The new market for set-top devices—at least *1 billion* set-top devices in the next decade—demands good performance *and* low cost.

But what does that have to do with personal computing? Plenty. That PowerPC 602 yields a set-top device that has the graphics performance equivalent to a Silicon Graphics Indigo workstation, yet will cost users only $200. Will people want to sit at their computer when they can find more computing power (and more network services) available on their TV?

The only way that a new software or hardware architecture can take over the desktop is when that desktop is undergoing rapid expansion. It happened that way after 1981, when 500,000 CP/M computers were replaced over a couple of years by more than 5 million MS-DOS computers. The market grew by an order of magnitude and all those new machines used new technology. But these days we're mainly replacing old machines, rather than expanding our user base, and most of the time we're using our new Pentium hardware to emulate 8086s at faster and faster speeds. That's not a prescription for revolution.

It's going to take another market expansion to drive a new software architecture, and since nearly every desk that can support a PC already has one sitting there, the expansion is going to have to happen where PCs *aren't*. The market expansion that I think is going to take place will be outside the office, to the millions of workers who don't have desks, and in the home, where computing has never really found a comfortable place. We're talking about something that's a cross between a television and a PC—the set-top box.

Still, a set-top device is not a computer, because it has no local storage and no software applications, right? Wrong. The set-top device will be connected to a network, and at the other end of that network will be companies that will be just as happy to download spreadsheet code as they are to send you a copy of *Gone with the Wind*. With faster network performance, a local hard disk is unnecessary. There goes the PC business. Also unnecessary is owning your own software, since it is probably cheaper to rent software that is already on the server. There goes the software business, too. But by converting, say, half the television watchers in the world into computer users, there will be 1 billion new users demanding software that runs on the new boxes. Here come whole new computer hardware and software industries—at least if Larry Ellison gets his way.

Ellison is the "other" PC billionaire, the founder of Oracle Systems, a maker of database software. The legendary Silcion Valley rake who once told me he wanted to be married "up to five days per week," has other conquests in mind. He wants to defeat Bill Gates.

"I think personal computers are ridiculous," said Ellison. "It's crazy for me to have this box on my desk into which I pour bits that I've brought home from the store in a cardboard box. I have to install the software, make it work, and back up my data if I want to save it from the inevitable hard disk crash. All this is stupid: it should be done for me.

"Why should I have to go to a store to buy software?" Ellison continued. "In a cardboard box is a stupid way to buy software. It's a box of bits and not only that, they are old bits. The software you buy at a store is hardly ever the latest release.

"Here's what I want," said Ellison. "I want a $500 device that sits on my desk. It has a display and memory but no hard or floppy disk drives. On the back it has just two ports—one for power and the other to connect to the network. When that network connection is made, the latest version of the operating system is automatically downloaded. My files are stored on a server somewhere and they are backed up every night by people paid to do just that. The data I get from the network is the latest, too, and I pay for it all through my phone bill because that's what the computer really is—an extension of my telephone. I can use it for computing, communicating, and entertainment. That's the personal computer I want and I want it *now!*"

Larry Ellison has a point. Personal computers probably are a transitional technology that will be replaced soon by servers and networks. Here we are wiring the world for Internet connections and yet we somehow expect to keep using our hard disk drives. Moving to the next standard of networking is what it will take to extend computing to the majority of citizens. Using a personal computer has to be made a lot easier if my mom is going to use a computer. The big question is when it all happens? How soon is soon? Well, the personal computer is already twenty years old, but my guess is it will look very different in another ten years.

Oracle, Ellison's company, wants to provide the software that links all those diskless PCs into the global network. He thinks Microsoft is so concentrated on the traditional stand-alone PC that Oracle can snatch ownership of the desktop software standard when this changeover takes place. It just might succeed.

If there's a good guy in the history of the personal computer, it must be Steve Wozniak, inventor of the Apple I and Apple II.

Prankster and dial-a-jokester, Woz was also the inventor of a pirated version of the VisiCalc spreadsheet called VisiCrook that not only defeated VisiCalc's copy protection scheme but ran five times faster than the original because of some bugs that he fixed along the way. Steve Wozniak is unique, and his vision of the future of personal computing is unique, too, and important. Wozniak is no longer in the computer industry. His work is now teaching computer skills to fifth- and sixth-grade students in the public schools of Los Gatos, California, where he lives. Woz teaches the classes and he funds the classes he teaches. Each student gets an Apple PowerBook 540C notebook computer, a printer, and an account on America Online, all paid for by the Apple cofounder. Right now, Woz estimates he has 100 students using his computers.

"If I can get them using the computer and show them there is more that they can do than just play games, that's all I want," said Wozniak. "Each year I find one or two kids who get it instantly and want to learn more and more about computers. Those are the kids like me and if I can help one of them to change the world, all my effort will have been worthwhile."

As a man who is now more a teacher than an engineer, Woz's view of the future takes the peculiar perspective of the computer educator trying to function in the modern world. Woz's concern is with Moore's Law, the very engine of the PC industry that has driven prices continually down and sales continually up. Woz, for one, can't wait for Moore's Law to be repealed.

Huh?

For the last thirty years and probably for another decade, Moore's Law will continue to apply. But while the rest of the computing world waits worriedly for that moment when the lines etched on silicon wafers get so thin that they are equal to the wavelength of the light that traces them—the technical dead end for photolithography—Steve Wozniak looks forward to it. "I can't wait," he said, "because that's when software tools can finally start to mature."

While the rest of us fear that the end of Moore's Law means the end of progress in computer design, Wozniak thinks it means the true coming of age for personal computers—a time to be celebrated. "In American schools today a textbook lasts ten years and a desk lasts twenty years, but a personal computer is obsolete when it is three years old," he said. "That's why schools can't afford computers for every child. And every time the computer changes, the software changes, too. That's crazy.

"If each personal computer could be used for twenty years, then the schools could have one PC for each kid. But that won't happen until Moore's Law is played out. Then the hardware architectures can stabilize and the software can as well. That's when personal computers will become really useful, because they will have to be tougher. They'll become appliances, which is what they should always have been."

To Woz, the personal computer of twenty years from now will be like haiku: there won't be any need to change the form, yet artists will still find plenty of room for expression within that form.

We overestimate change in the short term by supposing that dominant software architectures are going to change practically overnight, without an accompanying change in the installed hardware base. But we also underestimate change by not anticipating new uses for computers that will probably drive us overnight into a new type of hardware. It's the *texture* of the change that we can't anticipate. So when we finally get a PC in every home, it's more likely to be as a cellular phone with sophisticated computing ability thrown in almost as an after-thought, or it will be an ancillary function to a 64-bit Nintendo machine, because people need to communicate and be entertained, but they don't really need to compute.

Computing is a transitional technology. We don't compute to compute, we compute to design airplane wings, simulate oil fields, and calculate our taxes. We compute to plan businesses and then to understand why they failed. All these things, while parading as computing tasks, are really experiences. We can have enough power, but we can never have enough experience, which is why computing is beginning a transition from being a method of data processing to being a method of communication.

People care about people. We watch version after version of the same seven stories on television simply for that reason. More than 80 percent of our brains are devoted to processing visual information, because that's how we most directly perceive the world around us. In time, all this will be mirrored in new computing technologies. We're heading on a journey that will result, by the middle of the next decade, in there being no more phones or televisions or computers. Instead, there will be billions of devices that perform all three functions, and by doing so, will tie us all together and into the whole body of human knowledge. It's the next big wave, a veritable tsunami. Surf's up!

INDEX

INDEX

367